Problems and Solutions in Organometallic Chemistry

Susan E. Kegley
Middlebury College
Middlebury, Vermont
(currently Williams College
Williamstown, Massachusetts)

Allan R. Pinhas
University of Cincinnati
Cincinnati, Ohio

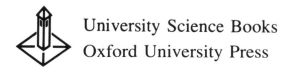
University Science Books
Oxford University Press

Oxford University Press, Walton Street, Oxford OX2 6DP

Oxford New York Toronto
Delhi Bombay Calcutta Madras Karachi
Petaling Jaya Singapore Hong Kong Tokyo
Nairobi Dar es Sallan Cape Town
Melbourne Auckland

and associated companies in
Beirut Berlin Ibadan Nicosia

Oxford is a trade mark of Oxford University Press

Published ing the United States
by University Science Books
Mill Valley, California

ISBN 0 19 855720 5 (OUP)

"If a man* can keep alert and imaginative, an error is a possibility, a chance at something new; to him wandering and wondering are part of the same process, and he is most mistaken, most in error, whenever he quits exploring."

William Least-Heat Moon, in "Blue Highways"

* or woman!

CONTENTS

PREFACE

Although quite a few textbooks dealing with transition metal organometallic chemistry have recently been published, there seems to be a general lack of problems to reinforce the principles discussed in these books. The aim of this book is to fill this void by providing real problems from the recent literature, complete with solutions and references. Additionally, we feel that it would be helpful to the practicing organometallic chemist to have a readily accessible compilation of IR and NMR spectral data for a variety of organometallic compounds, and to have descriptions of NMR techniques frequently used as tools in the spectroscopic characterization of organometallic compounds. Chapter 1 is such a source and Chapter 4 provides some practice problems related to dynamic NMR spectroscopy and fluxional processes. Chapter 2 deals with structure and bonding in organometallic compounds, while Chapters 3 and 5-8 cover the basic reaction types in organometallic chemistry: ligand substitution, oxidative-addition, reductive-elimination, migratory insertion, nucleophilic and electrophilic attack on coordinated ligands. Chapters 9, 10, and 11 on metallacycles, transition metal alkyls and hydrides, and catalytic reactions require the student to apply the principles learned in Chapters 3-8 to specific topics. Chapter 12 emphasizes the applications of organometallics to organic synthesis, and Chapter 13 is a collection of additional problems that require the student to use all facets of his or her knowledge of organometallic chemistry.

When designing problems, we chose to take a mechanistic approach. We feel that it is quite important to know why a reaction occurs and how to prove mechanistic hypotheses, rather than just know that the reaction does occur. For this reason, there are certain areas that are not covered as thoroughly as others, e.g. applications to organic synthesis and catalytic reactions, where for many systems, mechanistic information is largely unknown. We hope that exposure to a few of the known systems will be educational.

It should be kept in mind that for many problems, particularly those that require the student to propose experiments that would prove a mechanistic hypothesis, there is certainly more than one correct answer. The answers given in the book represent what has actually been done and published by workers in the field.

This book is intended to match the subject material covered in the second edition of Collman, Hegedus, Norton, and Finke's "Principles and Applications of Organotransition Metal Chemistry", thus the general chapter headings in the problem book are similar to those in the text; however, we feel that the problems in this book will be useful to any student of organometallic chemistry.

Susan E. Kegley

Allan R. Pinhas

ACKNOWLEDGMENTS

Let's face it—writing a book is a lot of trouble! But fortunately for us, there have been many people who have contributed their time, ideas, support, and hard work to help make it happen. We really appreciate their efforts. For helpful comments and discussions, we thank M. Brookhart, Bob Bergman, Jack Norton, Ken Whitmire, Dave Reingold, Jack Halpern, Jim Merrifield, and Steve Sontum. For communication of supplementary material and unpublished results, we thank M. Brookhart, Bob Bergman, John Bercaw, and John Gladysz. We thank Bill Lamanna for the use of some information from his Ph.D. thesis. For expert proofreading, we thank Joan Simunic. We would particularly like to express our appreciation to Jim Collman, Lou Hegedus, Jack Norton, and Rick Finke for sending preprints of the second edition of "Principles and Applications of Organotransition Metal Chemistry" to us. The title pages for chapters 1 and 4 were reprinted with permission from Organomet. **1986**, 5, 961 and J. Am. Chem. Soc. **1983**, 105, 464, respectively. MSD Isotopes was kind enough to allow us to reproduce their table "Deuterated NMR Solvents—Handy Reference Data". We would also like to acknowledge the support given by the Middlebury College Faculty Professional Development Fund.

We are extremely grateful to those who have assisted in the production of the camera-ready copy. For help with the drawings, we thank Chris Staats, Diane Copeland, Ebbe Hartz, Lisa Schmeichel, Brigid Nicholson, Gail Byers, and Jeff Kelly. The computer program to produce many of the drawings was skillfully written by Kipley Olson. For help with the typing and assembly of the manuscript, we thank Shevaun Mackie, Chris Staats, and Mrs. Joyce McAllister. The people at the Middlebury College computer center and in the Middlebury College Science Library were extremely helpful in this endeavor, particularly Tom Copeland and Sharon Strassner.

Finally, we are truly grateful for the infinite patience and understanding of our friends and co-workers Gail Byers, Kit Cummins, Ebbe Hartz, Shevaun Mackie, Brigid Nicholson, Chris Staats, Joan Simunic, Tim Figley, Nuan Chantarasiri, Pond Chamchaang, Dave Sullivan, and Pam Parente during the summer of 1986. We also feel strongly that Bruce Armbruster of University Science Books deserves a medal for his patience!

Thanks y'all!

ABBREVIATIONS

acac	acetylacetonate anion
AIBN	azoisobutyronitrile
Ar	aryl
n-Bu	n-butyl, $-CH_2CH_2CH_2CH_3$
t-Bu	t-butyl, $-C(CH_3)_3$
CIDNP	chemically induced dynamic nuclear polarization
cm^{-1}	wave number
COD	1,5-cyclooctadiene
COT	cyclooctatetraene
Cp	η^5-cyclopentadienyl, η^5-C_5H_5
Cp*	η^5-pentamethylcyclopentadienyl, η^5-C_5Me_5
d	doublet (NMR)
dd	doublet of doublets (NMR)
dt	doublet of triplets (NMR)
dq	doublet of quartets (NMR)
diphos	bis-1,2-diphenylphosphinoethane (dppe)
dmpe	bis-1,2-dimethylphosphinoethane
dppe	bis-1,2-diphenylphosphinoethane (diphos)
dtc^-	dithiocarbamate anion, $[(H_2N)CS_2]^-$
Et	ethyl, $-CH_2CH_3$
Fp	$CpFe(CO)_2$
HOMO	highest occupied molecular orbital
Hz	hertz, sec^{-1}
IR	infrared
L	a generic unidentate ligand
LUMO	lowest unoccupied molecular orbital

m	medium (IR), multiplet (NMR)
M	a generic metal
Me	methyl, $-CH_3$
MO	molecular orbital
NMR	nuclear magnetic resonance
^-OAc	acetate anion, $[OCOCH_3]^-$
^-OTf	triflate anion, $[OSO_2CF_3]^-$
PPN^+	bis(triphenylphosphine)iminium, $(Ph_3P)_2N^+$
Ph	phenyl, $-C_6H_5$
Ph_3P	triphenylphosphine, $P(C_6H_5)_3$
n-Pr	n-propyl, $-CH_2CH_2CH_3$
i-Pr	iso-propyl, $-CH(CH_3)_2$
p-tol	para-tolyl, $p-C_6H_4-CH_3$
q	quartet (NMR)
R	a generic alkyl group
rds	rate-determining step
RT	room temperature
s	strong (IR), singlet (NMR)
sh	shoulder (IR)
SST	spin saturation transfer
t	triplet (NMR)
THF	tetrahydrofuran, C_4H_8O
tos	toluenesulfonyl, $CH_3-C_6H_4-SO_2-$
TMS	trimethylsilyl, Me_3Si; tetramethylsilane, $SiMe_4$ (NMR)
UV	ultraviolet
w	weak (IR)
X	a halogen

Methods for the Identification of Organometallic Complexes

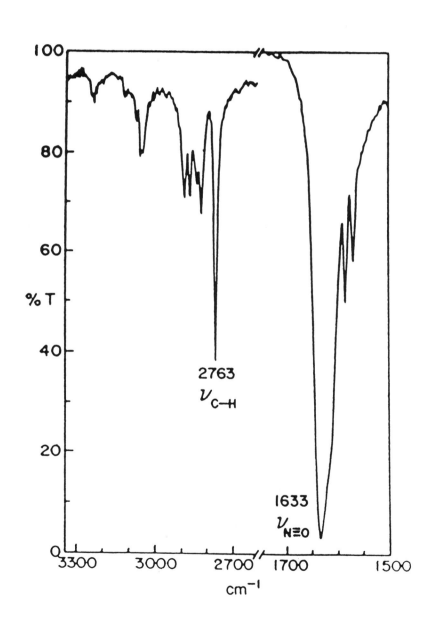

INTRODUCTION

A variety of techniques are now employed in the characterization of organometallic complexes, and it is therefore useful to have a general feel for the significance of the analytical data obtained by these techniques in the study of new complexes. Many methods are available for the characterization of organometallics, including nuclear magnetic resonance (NMR) spectroscopy, infrared (IR) spectroscopy, X-ray diffraction, neutron diffraction, mass spectroscopy, elemental analysis, electrochemistry, and UV-visible spectroscopy. Typically, the identification of a new organometallic compound involves the use of a combination of these techniques, with the most common being NMR spectroscopy and IR spectroscopy.

Definitive proof of structure is possible by the use of X-ray crystallographic analysis of a single crystal of the substance, accompanied by an elemental analysis. X-ray diffraction is a very powerful tool that can provide concrete evidence of atomic interactions that can only be hinted at by other analytical methods; however, it gives no information about any dynamic behavior of the molecule in solution, and it is limited by the requirement for single crystals and by the fact that the expertise and instrumentation are not always readily available. A collection of crystal structures of many organometallic compounds can be found in the series "Comprehensive Organometallic Chemistry" [1(a)].

Neutron diffraction [1(b)] is useful when it is necessary to specifically locate hydrogen atoms in a crystal structure. Typically, larger crystals are necessary for neutron diffraction than for X-ray diffraction, and there are only a few facilities in the U.S. where this analysis can be carried out, specifically at Argonne, Brookhaven, and Los Alamos National Laboratories.

Information about the electrochemical properties [2a,b] and the UV-visible spectrum of a compound is useful for specific applications, but these techniques are not routinely used for the characterization of organometallic complexes. Mass spectroscopic analysis [2c-f] is an important tool for determining the molecular weight of a compound, and knowledge of the fragmentation patterns for organometallic molecules can be of assistance in assigning a structure to a new compound.

The detailed information about molecular structure provided by NMR spectroscopy has made it a very popular technique and has led to a great demand for better and more easily used NMR spectrometers. This demand has been met successfully by manufacturers of NMR instrumentation and it is now common for a practicing organometallic chemist to have "hands-on" access to a high-field NMR spectrometer. Infrared spectrometers have long been readily available because of their inherent simplicity of design. This chapter will thus focus on the characterization of organometallic compounds by IR and NMR spectroscopy. The structure of a new organometallic compound can often be ascertained by comparison of its NMR and IR data to that of complexes of known structure, aided by a knowledge of the reaction chemistry of the particular complex; thus, it is useful to have a collection of spectral data with which to compare. The purpose of this chapter is to provide a central location for the practicing organometallic chemist to turn to when in pursuit of general trends for the NMR and IR spectral data of organometallic complexes. Additionally, there are a number of NMR techniques frequently used by organometallic chemists that will be described briefly, specifically variable temperature NMR and spin saturation transfer experiments as applied to the clarification of the fluxional behavior of organometallic molecules and the determination of the energy barriers involved. Although there are many interesting magnetically active nuclei, the contents of this chapter will be confined to carbon-13 and proton NMR data of mono- and dinuclear complexes. The reader is referred to other sources for information on other magnetically active nuclei (see Tables 1 and 2).

GENERAL INFORMATION: NMR SPECTROSCOPY

A number of excellent books have been written that present specific information on NMR and Fourier Transform (FT) NMR spectroscopy. Some are quite theoretical and very detailed [3], while others are written for the beginning NMR spectroscopist [4]. Books that address the more practical aspects of NMR, including hardware and software are also available [5]. In this chapter, it is assumed that the reader has a working knowledge of the basic principles of NMR spectroscopy and a feel for the location of organic proton and carbon resonances for strictly organic compounds.

There are many magnetically active nuclei that are relevant to organometallic chemistry, as listed in Tables 1 and 2 (data in these tables are taken largely from reference 5b, pp. 2-5). Table 1 contains the nuclei that have a spin quantum number (I) of 1/2, and therefore produce sharp lines in their NMR spectra. NMR spectra of nuclei with I > 1/2 (see Table 2) can also be obtained; however, because of higher nuclear spin quantum numbers and nuclear quadrupole moments, the NMR spectra of these elements typically produce broader lines than those nuclei with I = 1/2, with the amount of broadening dependent on the value of the quadrupole moment and the efficiency of quadrupolar relaxation for that particular nucleus. While NMR spectra of all magnetically active nuclei can be useful, the higher resolution obtainable for elements with I = 1/2 often makes NMR studies on molecules containing these elements more informative. More information on each nucleus is given in the references listed. The references are not meant to be comprehensive; instead, recent reviews are cited, which should provide a place to start looking for more information on NMR spectroscopy of that particular nucleus.

Practical Considerations

Solvents
When choosing a solvent for an NMR sample, several factors should be taken into consideration. The first and most obvious is that the solvent should dissolve the complex and not react with it. If variable temperature NMR spectroscopy is being carried out, the freezing point or boiling point of the solvent must be compatible with the temperatures to

TABLE 1: Magnetic Properties of Nuclei with I = 1/2

Isotope	Natural Abundance	Frequency(MHz) for a 23.5 kG field	Relative Sensitivity[a]	Reference
^{1}H	99.985	100.00	1.0	4-6
^{13}C	1.108	25.19	0.016	7
^{15}N	0.37	10.13	0.001	8
^{19}F	100.00	94.08	0.83	9
^{29}Si	4.67	19.86	0.08	10
^{31}P	100.00	40.48	0.07	11
^{57}Fe	2.24	3.24	0.000033	12
^{77}Se	7.50	19.10	0.007	13
^{89}Y	100.00	4.90	0.00012	14
^{103}Rh	100.00	3.15	0.000031	15
^{107}Ag	51.35	4.04	0.000067	16
^{109}Ag	48.65	4.65	0.0001	16
^{111}Cd	12.86	21.20	0.0095	17
^{113}Cd	12.34	22.18	0.011	17
^{117}Sn	7.67	35.62	0.045	18
^{119}Sn	8.68	37.27	0.052	18
^{123}Te	0.89	26.21	0.018	13
^{125}Te	7.03	31.59	0.032	13
^{183}W	14.58	4.11	0.00007	19
^{195}Pt	33.7	21.50	0.01	20
^{199}Hg	16.9	17.88	0.0057	20
^{203}Tl	29.5	57.22	0.19	21
^{205}Tl	70.5	57.79	0.19	21
^{207}Pb	21.1	20.86	0.01	22

[a] Relative sensitivity is scaled to ^{1}H = 1.

TABLE 2: Magnetic Properties of Nuclei with I > 1/2[a]

Isotope	Natural Abundance	I	Frequency(MHz) for a 23.5 kG field	Relative Sensitivity[b]	Reference
^{2}H	0.015	1	15.35	0.0096	23
^{10}B	18.83	3	10.74	0.02	24
^{11}B	81.17	3/2	32.08	0.165	24
^{14}N	99.64	1	7.22	0.001	25
^{17}O	0.037	5/2	13.56	0.029	26
^{27}Al	100.00	5/2	26.06	0.207	27
^{33}S	0.74	3/2	7.67	0.0023	22
^{45}Sc	100.00	7/2	24.29	0.301	20
^{47}Ti	7.75	5/2	5.63	0.0021	20
^{49}Ti	5.51	7/2	5.64	0.0038	20
^{51}V	99.76	7/2	26.29	0.38	20
^{53}Cr	9.54	3/2	5.65	0.0001	19
^{55}Mn	100.00	5/2	24.78	0.18	20
^{59}Co	100.00	7/2	23.73	0.28	28
^{61}Ni	1.25	3/2	8.90	0.0035	29
^{63}Cu	69.09	3/2	26.50	0.094	30
^{65}Cu	30.91	3/2	28.40	0.12	30
^{67}Zn	4.12	5/2	6.25	0.0029	20
^{69}Ga	60.2	3/2	24.00	0.069	29
^{71}Ga	39.8	3/2	30.50	0.142	29
^{73}Ge	7.61	9/2	3.49	0.0014	29
^{75}As	100.00	3/2	17.13	0.025	29
^{91}Zr	11.23	5/2	9.30	0.0094	29
^{93}Nb	100.00	9/2	24.44	0.482	20

(Continued on following page)

TABLE 2. (cont.)

Isotope	Natural Abundance	I	Frequency(MHz) for a 23.5 kG field	Relative Sensitivity[b]	Reference
^{95}Mo	15.78	5/2	6.51	0.0032	19
^{97}Mo	9.60	5/2	6.65	0.0034	19
^{99}Tc	100.00	9/2	4.61	0.4	29
^{99}Ru	12.81	5/2	22.51	0.0011	22
^{101}Ru	16.98	5/2	5.17	0.0014	22
^{105}Pd	22.23	5/2	4.58	0.00078	29
^{113}In	4.16	9/2	21.87	0.345	29
^{121}Sb	57.25	5/2	23.93	0.16	29
^{123}Sb	42.75	3/2	12.96	0.046	29
^{139}La	99.91	7/2	14.12	0.059	20
^{177}Hf	13.39	7/2	4.00	0.00064	29
^{179}Hf	13.78	9/2	2.52	0.0022	29
^{181}Ta	100.00	7/2	11.99	0.036	20
^{185}Re	37.07	5/2	22.52	0.13	20
^{187}Re	62.93	5/2	22.74	0.14	20
^{189}Os	16.1	3/2	7.76	0.0022	29
^{191}Ir	38.5	3/2	1.72	0.000035	29
^{193}Ir	61.5	3/2	1.87	0.000042	29
^{197}Au	100.00	3/2	1.72	0.000025	29
^{209}Bi	100.00	9/2	16.07	0.14	29

[a]This table does not include data for Group I and II metals, the halogens, the noble gases, the lanthanides or the actinides.

[b] Relative sensitivity is scaled to ^{1}H = 1.

be encountered in the experiment. For most organometallic applications, dry, deoxygenated solvents are essential. When comparing the chemical shifts of an unknown complex to those of a literature compound, it is important to note the solvent used in both situations. Chemical shifts and coupling constants are quite solvent-dependent, and large changes in both are often observed, particularly between aromatic and non-aromatic solvents. Table 3 (courtesy of MSD Isotopes) lists commonly available deuterated solvents with their physical and spectral properties.

Sample Preparation

When preparing a sample, great care should be taken to insure that the solution is homogeneous and does not contain bits of paramagnetic material which will decrease magnetic relaxation times and therefore produce broad lines in the spectrum (soluble paramagnetic materials will also cause this problem). A method for removing insoluble paramagnetic impurities from the solution is to centrifuge the sample for a minute or two to force the undissolved solids to the bottom (or the top in a sealed sample) of the tube. When variable temperature work is to be done on a sample, the solvent should be degassed and the NMR tube should be flame sealed.

Potential Problems

Because organometallic complexes contain carbons and hydrogens in environments not encountered in purely organic compounds, the range of chemical shifts is somewhat larger for organometallic complexes. The result of this is that a proton or carbon resonance might easily be missed if a standard sweep width is used (0-12 ppm for protons, 0-220 ppm for carbons). For example, the chemical shift of a hydrogen directly bound to a transition metal may be found as far upfield as -40 ppm, or that of hydrogens bound to a transition metal carbene carbon may be found as far downfield as 20 ppm. In the carbon-13 NMR spectra of organometallic complexes, it is not unusual to find carbonyl carbons at 250 ppm, carbene carbons at 350 ppm, and carbons directly bound to the metal at -40 ppm. "Folded back" resonances are a good indication of protons or carbons that are not within the spectral range being observed. These "fold-back" resonances are observed only in FT-NMR spectra and usually appear as unphasable peaks that are (unfortunately) quite easy to dismiss as an instrumental "glitch". When this occurs or when a resonance is expected at an unusual location, it is necessary to widen the sweep width, i.e., increase the radio frequency "window" being observed.

Another problem that frequently arises is inaccuracy of the integrals obtained for the proton NMR of many organometallic complexes. Generally, this is due to long spin-lattice relaxation times (T_1's) for protons in certain environments. In an FT experiment, insufficient relaxation of the nuclei between pulses results in saturation of the nuclei and excitation of fewer nuclei for each successive pulse. The result is that the nuclei with long T_1's will produce a weaker signal relative to that observed for nuclei with short T_1's, and the integrals will be inaccurate (See reference 3(c) or 4(a) for a brief explanation of T_1's). Cyclopentadienyl hydrogens are particularly prone to this problem.

TABLE 3[a]

DEUTERATED NMR SOLVENTS-HANDY REFERENCE DATA

Compound Mol. Wt.	d_4^{20}	m.p.*	b.p.*	δ_H (mult)[b]	J_{HD}	δ_C (mult)[b]	J_{CD} (J_{CF})
Acetic Acid-d₄ 64.078	1.12	17	118	11.53 (1) 2.03 (5)	2	178.4 (br) 20.0 (7)	20
Acetone-d₆ 64.117	0.87	-94	57	2.04 (5)	2.2	206.0 (13) 29.8 (7)	0.9 20
Acetonitrile-d₃ 44.071	0.84	-45	82	1.93 (5)	2.5	118.2 (br) 1.3 (7)	21
Benzene-d₆ 84.152	0.95	5	80	7.15 (br)		128.0 (3)	24
Chloroform-d 120.384	1.50	-64	62	7.24 (1)		77.0 (3)	32
Cyclohexane-d₁₂ 96.236	0.89	6	81	1.38 (br)		26.4 (5)	19
Deuterium Oxide 20.028	1.11	3.8	101.4	4.63 (DSS) 4.67 (TSP)			
1,2-Dichloroethane-d₄ 102.985	1.25	-40	84	3.72 (br)		43.6 (5)	23.5
Diethyl-d₁₀ Ether 84.185	0.82	-116	35	3.34 (m) 1.07 (m)		65.3 (5) 14.5 (7)	21 19
Diglyme-d₁₄ 148.263	0.95	-68	162	3.49 (br) 3.40 (br) 3.22 (5)	1.5	70.7 (5) 70.0 (5) 57.7 (7)	21 21 21
Dimethylformamide-d₇ 80.138	1.04	-61	153	8.01 (br) 2.91 (5) 2.74 (5)	2 2	162.7 (3) 35.2 (7) 30.1 (7)	30 21 21
Dimethyl-d₆ Sulphoxide 84.170	1.18	18	189	2.49 (5)	1.7	39.5 (7)	21
p-Dioxane-d₈ 96.156	1.13	12	101	3.53 (m)		66.5 (5)	22
Ethyl Alcohol-d₆ (anh) 52.106	0.91	<-130	79	5.19 (1) 3.55 (br) 1.11 (m)		56.8 (5) 17.2 (7)	22 19
Glyme-d₁₀ 100.184	0.86	-58	83	3.40 (m) 3.22 (5)	1.6	71.7 (5) 57.8 (7)	21 21
Hexafluoroacetone Deuterate[c] 198.067	1.71	21		5.26 (1)		122.5 (4) 92.9 (7)	(287) (34.5)
HMPT-d₁₈ 197.314	1.14	7	106 (11)	2.53 (2 x 5)	2 (9.5)	35.8 (7)	21
Methyl Alcohol-d₄ 36.067	0.89	-98	65	4.78 (1) 3.30 (5)	1.7	49.0 (7)	21.5
Methylene Chloride-d₂ 86.945	1.35	-95	40	5.32 (3)	1	53.8 (5)	27
Nitrobenzene-d₅ 128.143	1.25	6	211	8.11 (br) 7.67 (br) 7.50 (br)		148.6 (1) 134.8 (3) 129.5 (3) 123.5 (3)	24.5 (p) 25 26
Nitromethane-d₃ 64.059	1.20	-29	101	4.33 (5)	2	62.8 (7)	22
isoPropyl Alcohol-d₈ 68.146	0.90	-86	83	5.12 (1) 3.89 (br) 1.10 (br)		62.9 (3) 24.2 (7)	21.5 19
Pyridine-d₅ 84.133	1.05	-42	116	8.71 (br) 7.55 (br) 7.19 (br)		149.9 (3) 135.5 (3) 123.5 (3)	27.5 24.5 (γ) 25
Tetrahydrofuran-d₈ 80.157	0.99	-109	66	3.58 (br) 1.73 (br)		67.4 (5) 25.3 (br)	22 20.5
Toluene-d₈ 100.191	0.94	-95	111	7.09 (m) 7.00 (br) 6.98 (m) 2.09 (5)	2.3	137.5 (1) 128.9 (3) 128.0 (3) 125.2 (3) 20.4 (7)	23 24 24 (p) 19
Trifluoroacetic Acid-d[d] 115.030	1.50	-15	72	11.50 (1)		164.2 (4) 116.6 (4)	(44) (283)
2,2,2-Trifluoroethyl Alcohol-d₃[e] 103.059	1.45	-44	75	5.02 (1) 3.88 (4 x 3)	2 (9)	126.3 (4) 61.5 (4 x 5)	(277) 22 (36)

*Melting and boiling points (in °C) are those of the corresponding light compound (except for D₂O) and are intended only to indicate the useful liquid range of the materials.

[b]¹H (of the residual protons) and ¹³C spectra were determined on HA-100 and XL-100-15 spectrometers, respectively, for the same sample of each solvent containing 5% TMS (v/v). The chemical shifts are in ppm relative to TMS; the coupling constants are in Hz. (Since deuterium has a spin of 1, triplets arising from coupling to deuterium have the intensity ratio of 1:1:1, etc.) The multiplicity br indicates a broad peak without resolvable fine structure, while m denotes one with fine structure. It should be noted that the chemical shifts, in particular, can be dependent on solute, concentration and temperature.

[c]δ_F (CFCl₃) 82.6 (1)   [d]δ_F (CFCl₃) 76.2 (1)   [e]δ_F (CFCl₃) 77.8 (5), J_{FD} 1.2   all determined on an HA-100 spectrometer.

Left margin:
P.O. BOX 2951 TERMINAL ANNEX
LOS ANGELES, CA 90051
213 723-9521

800 325-9034

MERCK & CO., Inc.
4545 OLEATHA AVE.
ST. LOUIS, MO 63116
314 353-7000
ISOTOPES

P.O. BOX 899
POINTE CLAIRE-DORVAL
QUEBEC H9R 4P7
514 697-2823

MERCK
SHARP
& DOHME
ISOTOPES

[a]Reprinted with permission from MSD Isotopes

There are several solutions to this problem, one of which involves using a smaller pulse width so that fewer nuclei are excited to the upper spin state with each pulse. Alternatively, inserting a delay between pulses (5-20 seconds for protons, up to several minutes for carbons) will allow time for relaxation to occur. A combination of these two techniques is frequently the best compromise in terms of instrument time required to obtain a good spectrum.

Special Considerations for Carbon-13 NMR

Long relaxation times present a special problem when obtaining carbon-13 NMR spectra of metal carbonyl compounds. Indeed, it is not uncommon for M-CO resonances to be completely invisible under normal spectral acquisition conditions. The tricks used above to obtain good integrals for proton NMR (using a smaller pulse width and inserting a pulse delay) are often not enough to facilitate obtaining a carbon spectrum within a reasonable period of time; however, there are other alternatives. Perhaps the most effective one is the use of "shiftless" relaxation reagents such as $Cr(acac)_3$. These reagents are paramagnetic species that provide a pathway for rapid relaxation of excited nuclei and thus enhance the signal intensity without shifting the position of the resonances significantly. Because these reagents are paramagnetic, the carbons that are part of the reagent do not show up in the spectrum. Addition of 1% $Cr(acac)_3$ is sufficient to provide effective relaxation for the carbonyl carbons of an organometallic complex [31]. Obtaining the spectrum at low temperatures also serves to decrease the T_1's of the carbons of interest and therefore increase the signal intensity [32]. This method works because the ability of the molecule to relax increases as the molecular correlation time (τ_c, the average time it takes for a molecule to reorient in solution) increases. At lower temperatures, τ_c increases due to a decrease in energy available and to increased solvent viscosity, and thus T_1 decreases at lower temperatures; i.e., $T_1 \propto 1/\tau_c$. (For more information on correlation times, see reference 4(a), pp. 116-117 and pp. 128-129).

Coupling Constants

The value of a proton-proton coupling constant gives information about the relative spatial orientation of the two protons. In aliphatic systems, the value of the coupling constant is proportional to the dihedral angle (ϕ) between the two C-H bonds. The mathematical relationship is given by the Karplus equation:

$$J = 10 \cos^2 \phi$$

For example, when the dihedral angle is $180°$ as for Ha and Hb, the value of the coupling constant will be ca. 10 Hz. When $\phi = 60°$, as for Ha and Hc, J should be ca. 2.5 Hz and when $\phi = 90°$, J should be zero.

For alkenes, the value of the coupling constant is indicative of the relative stereochemical orientation (i.e., cis, trans, or geminal) of the coupling protons. Cis couplings are usually small (3-18 Hz), trans couplings are larger (12-24 Hz) and geminal couplings are very small (-3 to 7 Hz) [4(a)].

Often there is more than one proton coupling to the proton in question and the result is a complex spectrum that may be difficult to interpret. In many cases, simplification of this type of spectrum is possible by a simple homonuclear decoupling experiment [4(a)] or if experimentally possible, by substitution of deuterium for one of the coupling hydrogens. For extremely complex systems, simplification of the spectrum can be achieved by computer simulation or through the use of two dimensional (2D) NMR techniques [33]. Some recent examples of 2D NMR applied to organometallic systems are cited in reference 34.

Carbon-carbon couplings are usually not observed in samples containing only natural abundance carbon-13, since the probability of having two carbon-13 atoms adjacent to one another is very small; however, carbon-hydrogen couplings are readily observable in the carbon spectrum acquired with off-resonance decoupling of the protons (see reference 4(a) for a description of off-resonance decoupling). The value of the coupling constant is proportional to the percent s character (ρ) of the C-H bond as given by the equation below, and thus the hybridization of the carbon atoms in a molecule can readily be determined.

$$^1J_{CH} = 500\,\rho$$

This relationship can only be strictly applied to simple hydrocarbons; however, an unexpected C-H coupling constant can be indicative of unusual bonding situations which can be further clarified by other means.

Coupling to other magnetically-active nuclei should also be anticipated. With nuclei having $I = 1/2$ (see Table 1), proton-like splittings will be observed; however, the coupling constants may be quite large compared to C-H or H-H couplings. For carbons or hydrogens bound to nuclei with $I > 1/2$ (see Table 2), quadrupolar broadening may be observed. If this interferes with resolution, acquisition of the spectrum at a lower temperature will often decrease quadrupolar broadening.

IDENTIFICATION OF ORGANOMETALLIC COMPLEXES

What follows is a number of tables of general information on typical proton and carbon-13 chemical shifts and IR stretching frequencies for commonly-encountered organometallic ligands. Where possible, the spectral data for the free ligands have also been included. These tables are meant to be a general guide. There certainly will be some situations in which the spectral characteristics of a complex will not fall in the range given. For Tables 6 and 7, the IR bands listed are for stretching frequencies only. Keep in mind that for NMR spectra, "low field" means larger δ values and "high field" means smaller δ values (in some cases, even negative).

TABLE 4. Spectral Data for Uncomplexed Unsaturated Organic Ligands

Ligand	^{1}H NMR(δ, ppm)[a]	^{13}C NMR(δ, ppm)[a]	Reference
C_2H_4	5.33(s)	123.3	7(f)
C_2H_2	1.80(s)	71.62 (C_6D_6)	35
	H1: 5.90(m)[b] H2: 1.5-3.4(m) H3: 1.27(m) H4: 1.50(m) H5: 5.41(br, s)	c	36
	H1: 2.80(m)[b,d] H2: 6.28(m) H3: 6.42(m)	C1: 42.2 C2: 133.0 C3: 133.4	37
	CH_3(1): 0.95(d,J=8 Hz) CH_3(2,3): 1.75(br,s) H: 2.4(m)	C1: 52.08[g] C2: 134.14 C3: 137.17 CH_3(1): 14.56 CH_3(2): 11.29 CH_3(3): 11.78	38
$MgCl^+[C_5Me_5]^-$	c	ring C: 108.51[g] CH_3: 11.42	38
$Na^+[C_5H_5]^-$	5.50(s) (THF) 5.25(s) (DMSO)	102.1 (C_6D_6)	37
$Tl^+[C_5H_5]^-$	6.08(s) (DMSO)	107.5 (C_6D_6)	37,7(f)
	c	C1: 113.1 C2: 103.1 C3: 101.7 CH_3: 14.4	7(f)
C_6H_6	7.15(s) (C_6D_6)	128.0	Table 3
	H1: 2.20(m) H2: 5.28(m) H3: 6.12(m) H4: 6.55(m)	C1: 28.1 C2: 120.4 C3: 126.8 C4: 131.0	39

(Continued on the following page)

Table 4. (cont.)

Ligand	^{1}H NMR (δ, ppm)[a]	^{13}C NMR (δ, ppm)[a]	Reference
$[C_7H_7]^+$	9.55[e]	156.2[f]	40
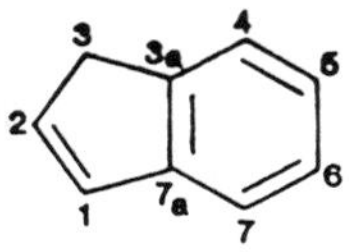	H1: 6.67(d, 5.6 Hz)[d] H2: 6.14(dd) H3: 2.99(d, J=2 Hz) H4-7: 6.5-7.5(m)	C1: 133.6 C2: 131.8 C3: 38.7 C3a: 143.3 C4: 123.4 C5: 126.0 C6: 124.2 C7: 120.6 C7a: 144.6	37,41

[a] Spectra were taken in $CDCl_3$ or CCl_4 unless otherwise noted.

[b] Cyclopentadiene exists as the dimer unless freshly distilled.

[c] Not found.

[d] Reference did not list solvent.

[e] PF_6^- salt in acetone-d_6.

[f] ClO_4^- salt in CD_3CN.

[g] In THF-d_8.

13

TABLE 5. Spectral Data for Unsaturated Organic Ligands Bound to Transition Metals

Ligand	^1H NMR (δ, ppm)[a]	^{13}C NMR (δ, ppm)[a,b]	Reference
η^2-C_2RR'	6–14[c]	110–230[d,e]	7(f),42
μ^2-(C_2RR')	4–6[c]	60–115	7(f),43
η^2-C_2H_4	0.5–5[f]	7–110[a,f]	7(f),44
$[\eta^3$-allyl$]^-$	terminal H: 2–4[g] central H: 4–6	terminal C: 37–80[h,i] central C: 88–125	7(f)
η^4-C_4R_4	1.5–6[c,f]	60–110	7(f),45
$[\eta^5$-$C_5H_5]^-$	3.5–7	75–123[j]	6,7(f)
$[\eta^5$-$C_5Me_5]^-$	1.5–2.0	ring C: 88–125 CH_3: 0–20	6,7(f)
	CH_3: 1.9–2.5 H1,H2: 3.5–6.5	83–140[k]	6,7(f)
η^6-C_6H_6	4–7.0	74–111	6,7(f)
	H1: 2.0–3.0 H2: 2.5–4 H3: 4–5 H4: 5–6	C1: 25–30 C2: 49–65 C3,C4: 90–106	7(f),46
$[\eta^7$-$C_7H_7]^+$	4.5–6.5[l]	80–105[l]	7(f),46
	H1,H3: 5–7 H2: 5–6.5 H4-7: 6–8	C1,C3: 60–115 C2: 65–110 C3a,C7a: 70–135[m] C4,C7: 120–130 C5,C6: 110–125	7(f),41

[a] Strongly electron-donating or electron-withdrawing substituents may shift the observed resonances out of the given range.

[b] See reference 7(f) for an extensive collection of ^{13}C NMR data for specific complexes.

(Notes continued on following page)

14

Notes for TABLE 5 (cont.)

[c]For R or R' = H.

[d]Acetylene can act as a two-, three-, or four-electron donor. The chemical shifts of the acetylenic carbons have been correlated with the number of electrons donated to the metal, with the lower field shifts observed for acetylenes donating four electrons and the higher field shifts observed for acetylenes donating two electrons. Intermediate values are observed for acetylenes acting as three-electron donors [42(b)].

[e]Platinum-acetylene complexes have chemical shifts for the acetylenic carbons over the range 70-115 ppm (see reference 7(f), p. 198).

[f]An extremely wide range of chemical shifts is observed for these types of complexes. See reference 7(f) for specific examples for comparison.

[g]Terminal allyl protons may resonate at unusually high field (0-1 ppm).

[h]Varies greatly with substituents.

[i]Sometimes observed out of range at lower field when substituents are present.

[j]Most Cp resonances seem to fall in this range, with the exception of some osmium complexes which have δCp around 65 ppm (see reference 7(f), p. 238).

[k]The ring carbon bound to the methyl group generally resonates at lower field than the other ring carbons.

[l]Most η^7-C_7H_7 resonances (both ^{13}C and ^{1}H) fall in the range given, with the exception of some bi- and trinuclear mixed-metal compounds (see reference 7(f), p. 258).

[m]The position of the chemical shifts of these carbons has been correlated to the hapticity of the indenyl ligand [41(b)]. For pentahapto indenyl complexes, carbons C3a and C7a fall at the low end of this range, and for trihapto complexes, carbons C3a and C7a fall at the high end of this range. A more quantitative evaluation of carbon-13 chemical shifts and hapticity of the indenyl ligand bound to d^6 and d^8 metals has also been made [41(a)].

TABLE 6. η^1-Ligands Bound to Transition Metals

Ligand	^{1}H NMR(δ, ppm)	^{13}C NMR(δ, ppm)	IR(cm^{-1})	Reference
L_nM-H	-24 to -7^a $2-4^b$	——	1900–2250	47
free CO	——	183.4	2143	47,48
L_nM-CO	——	180–250	1750–2100	7(f),47
$\mu^2-(CO)M_2L_n$	——	220–310^c	1700–1850	7(f),47
free CS	——	d	1274^e	47
L_nM-CS	——	295–360	1160–1410	49
L_nM-CR_3	-2 to 3^f	-50 to 113	g	7(f)
$L_nM=CHR$	8–20	200–450	h	7(f)
$\mu^2-CRH-M_2L_n$	5–12^i 1.4–2.6^j	100–210	g	50
$L_nM{\equiv}C-R$	k	230–365	3–6	7(f),50(b)
$L_nM-CHROH$	OH: 3–5^l C$\underline{H}$R: 4–7	65–90	3200–3600	51
$L_nM\overset{\displaystyle O}{\overset{\|}{-}}C-R$	2–5	190–300	1620–1680	7(f),47
$L_nM\overset{\displaystyle O}{\overset{\|}{-}}C-H$	12–17	240–310	1530–1630	52
Free N_2	——	——	2331^m	53
L_nM-N_2	——	——	1900–2200	47,54
NO(g)	——	——	1876	53
NO^+	——	——	2150–2400^n	55
L_nM-NO	——	——	2050–1450°	56
$\mu^2-(NO)M_2L_n$	——	——	1450–1550	57
$L_nM=O$	——	——	825–1100	58
M–O–M	——	——	700–900	58
Free O_2	——	——	1555^m	53
L_nM-O_2	——	——	1100–1200	59

[a]Hydrides bridging two or more metals generally resonate at higher field than terminal hydrides, even as high as -40 ppm.

(Notes continued on following page)

[b] For d^0 and d^{10} metal hydrides [60].

[c] The chemical shift ranges for bridging and terminal CO's overlap; however, if both types of CO's are present in the same compound (a frequent occurrence), the μ-CO will be observed at significantly lower field (20–70 ppm) than the terminal CO.

[d] Not stable.

[e] Isolated in a CS_2 matrix [47].

[f] For R=H.

[g] Nothing characteristic.

[h] Not found.

[i] For complexes with a M—M bond.

[j] For complexes without a M—M bond.

[k] Not found for R=H.

[l] Most —OH resonances of hydroxymethyl complexes occur in this range; however, one such complex, $Cp^*Ru(CO)_2CH_2OH$ was reported to have a shift of 0.88 for this proton.

[m] Raman active.

[n] Depending on counterion.

[o] The nitrosyl group can be bound in two ways, linear or "bent". Generally, bent nitrosyl groups have IR stretching frequencies between 1525 and 1690 cm^{-1}, i.e., they occur at the lower end of the range; however, some linear NO's also have NO stretching frequencies in the same range, so this cannot be used as a definitive criterion for identification of bent vs. linear NO.

TABLE 7. η^2-Ligands Bound to Transition Metals

Ligand	^{1}H NMR(δ, ppm)	^{13}C NMR(δ, ppm)	IR(cm^{-1})	Reference
Free C_2H_4	5.33[a]	123.3[a]	1623[b]	7(f),53
η^2-C_2H_4	0.5–5[c]	7–110[c]	1500–1600[b]	7(f),44,53
Free C_2RR'	1.8–3.5[d]	60–90	2190–2260[e]	35(a),53
η^2-C_2RR'	6–14[d]	110–230[f,g]	1700–2000[e]	7(f),42
η^2-acyl	2–5	190–300	1465–1620	61
free H_2	4.62	---	4161	62,53
η^2-H_2	–3 to –13[h]	---	2690[i]	63
L_nM (C/H triangle)	–16 to 5[j]	j,k	2350–2700	64
η^2-O_2	---	---	800–900	53

[a] In $CDCl_3$.

[b] C=C stretch.

[c] An extremely wide range of chemical shifts is observed for these types of complexes. See reference 7(f) for specific examples for comparison.

[d] For R or R'= H.

[e] C≡C stretch.

[f] Acetylene can act as a two-, three-, or four-electron donor. The chemical shifts of the acetylenic carbons have been correlated with the number of electrons donated to the metal, with the lower field shifts observed for acetylenes donating four electrons and the higher field shifts for acetylenes donating two electrons. Intermediate values are observed for acetylenes acting as three-electron donors [42(b)].

[g] Platinum-acetylene complexes have chemical shifts for the acetylenic carbons over the range 70–115 ppm (see reference 7(f), p. 198).

[h] These complexes are somewhat difficult to characterize definitively without X-ray or (preferably) neutron diffraction; however, Crabtree [63a–d] has recently discovered that the spin lattice relaxation time (T_1) of the hydrogens in the η^2-H_2 ligand are significantly shorter (15–40ms) than those of classical hydrides (390–800 ms) in the examples studied. The polyhdride examples of dihydrogen complexes often exhibit exchange between the M–H and M–H_2 units, which can lead to observation of short T_1's for the classical hydrides. As a result, it may be necessary to cool the sample to slow down the fluxional process in order to get accurate T_1 values.

(Notes continued on following page)

Notes for TABLE 7 (cont.)

[i] Generally weak or unobserved.

[j] At room temperature, these complexes are often fluxional between an alkyl-hydride complex and a 16-electron species with the (once-bridging) hydrogen bound only to the carbon. If this is so, the hydride resonance may be broad and will be located at a weighted average of the two possible static structures. The $^1J_{C-H}$ is also an average of that observed for the two possible structures.

[k] In the same range as normal alkyl carbons; however, the $^1J_{C-H}$ coupling constant for this type of complex is unusually low (75-100 Hz), due to the weakened C-H bond.

DYNAMIC NUCLEAR MAGNETIC RESONANCE

The structures of many organometallic molecules in solution at room temperature are not static, and frequently this fluxional behavior can give important information about structure, reactivity, and reaction mechanisms. These dynamic processes can be studied by NMR spectroscopy. A great deal is presently known about the theory and practice of dynamic NMR (DNMR) as applied to organometallic systems [65], and the purpose of this section is to introduce the student to the types of information obtainable from a DNMR experiment.

Dynamic processes can be classified into two categories, intramolecular or intermolecular. Some examples of intramolecular processes include skeletal rearrangements of ligands around a metal, rotation around metal-metal or metal-ligand bonds (e.g., rotation around a metal-olefin bond), "ring-whizzing" [65(b)], hapticity changes for unsaturated organic ligands, and interconversion of bridging and terminal ligands such as CO, CH_3, or NO. Intermolecular processes generally involve ligand dissociation/reassociation. If the exchanging nuclei are coupled to another magnetically active nucleus in the molecule, the couplings are retained if no bonds between the magnetically active nuclei are broken. In this situtation, if rapid exchange is occurring, the value of the coupling constant is an average of the individual coupling constants for the nuclei in the different environments. If a bond between the magnetically active nuclei is broken during the exchange process, coupling to the other nucleus is lost.

Broad lines in the NMR spectrum of a molecule are one indication that a fluxional process is occurring. The lineshapes and positions will change with temperature in a way that is governed by the particular process that is exchanging the two sites and by the rate constants associated with this process. These first-order or pseudo-first-order rate constants give a measure of the lifetimes of species in equilibrium. Two different methods of obtaining these rate constants from NMR spectra will be discussed, line shape analysis and spin saturation transfer (SST). In general, line shape analysis of proton NMR spectra can be used to obtain rate constants for processes with rate constants between 10^0 to 10^3 s^{-1}, while spin saturation transfer can be used to obtain rate constants for processes that are an order of magnitude slower, down to 10^{-1} s^{-1}.

Line Shape Analysis

Figure 1-1 shows the broad-band-deuterium-decoupled, temperature-dependent proton NMR spectrum of cyclohexane-d_{11} [66]. This is the simplest of exchanging systems, having only two sites (axial and equatorial), with no coupling present, and with equal populations of nuclei at the two sites.

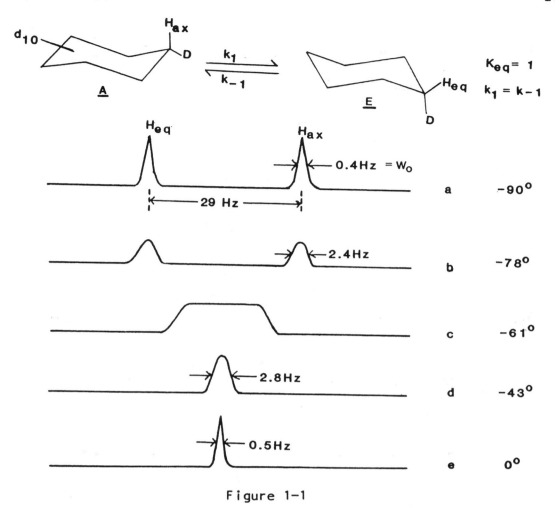

Figure 1-1

The spectrum at the slow exchange limit (**a**, T = -90°C and below) demonstrates that the lifetime of both **A** and **E** is long relative to the timescale of the NMR experiment, i.e., the exchange between axial and equatorial positions is slow. Under these conditions, the linewidth of the peak is governed by the spin-spin relaxation time, T_2, and any viscosity broadening that might exist at low temperatures. This linewidth is denoted by W_o and is measured as the linewidth at half-height.

As the sample is warmed, the two distinct peaks for H_a and H_e begin to broaden and merge (**b**), and at the coalescence temperature, T_c = -61°C (**c**), they coalesce into a single broad peak with a flat top. On further warming, the single peak sharpens (**e**). For this specific example, the system is said to be at the fast exchange limit at temperatures above 0°C, where the lifetimes of both species **A** and **E** are short relative to the timescale of the NMR experiment. The position of the peak at the fast exchange limit, δ_f is a weighted average of the positions of the two peaks visible for the static structure, as given by equation (1),

$$\delta_f = p_A \delta_A + p_E \delta_E \tag{1}$$

where p_A and p_E are the fractional populations of nuclei in sites A and E, and δ_A and δ_E are the chemical shifts of the axial and equatorial protons respectively. Since the two sites have equal populations, p_A and p_E are both 0.5, and the (now rapidly-exchanging) hydrogens will have a chemical shift exactly between the shifts observed for the two different hydrogens at the slow exchange limit.

There are several important mathematical relationships that apply at various temperatures. At the slow exchange limit (**a**),

$$k \ll \nu_A - \nu_E \tag{2}$$

where ν_A is the chemical shift (in Hz) of the axial proton and ν_E is the chemical shift of the equatorial proton. An approximation for obtaining rate constants at temperatures where slow exchange is taking place (**b**) involves using equation (3).

$$k = \pi (\Delta W) \tag{3}$$

where $\Delta W = W - W_o$, and W is the width at half-height of the broadened peak. For the cyclohexane example at -78°C

$$k(-78^\circ C) = \pi(2 \ s^{-1}) = 6.28 \ s^{-1} \tag{4}$$

At the coalescence temperature (**c**), the rate constant for exchange can be obtained from equation (5).

$$k = \frac{\pi (\nu_A - \nu_E)}{\sqrt{2}} \tag{5}$$

and for cyclohexane

$$k(-61^\circ C) = \frac{\pi(29 \ s^{-1})}{\sqrt{2}} = 64 \ s^{-1} \tag{6}$$

It is interesting to note that the greater the difference in Hz of the chemical shifts of the two exchanging protons, the faster the exchange rate required for coalescence to occur, and the higher T_c.

After coalescence (**d**), the rate constant can be calculated using the fast exchange approximation given by equation (7),

$$k = \frac{\pi (\nu_A - \nu_E)^2}{2(W - W_f)} \tag{7}$$

where W is again the width of the broadened peak at the temperature in question and W_f is the width at half height of the peak at the fast exchange limit. For cyclohexane,

$$k(-43^\circ C) = \frac{\pi (29 \text{ s}^{-1})^2}{2(2.3 \text{ s}^{-1})} = 566 \text{ s}^{-1} \tag{8}$$

At the fast exchange limit, e,

$$k \gg \nu_A - \nu_E \tag{9}$$

The two-spin, equal population example is the simplest system and is the only one that will be discussed here. The analysis can become very complex if the two sites are not equally populated and/or if coupling to other nuclei exists. The reader is referred to other sources [65] for discussion of more involved systems.

Several experimental aspects must also be taken into consideration when using or obtaining rate data from line shape analysis.

1) It is important to realize that the coalescence temperatures obtained in this manner are field-dependent, i.e., the $\Delta\nu$ is given in Hz, not ppm; therefore, any report of a coalescence temperature must be accompanied by the field strength of the magnet used in the experiment.

2) It is important to be aware that the rate of a fluxional process is frequently solvent-dependent [67].

3) Precise temperature measurement and control of temperature fluctuations are essential for obtaining accurate rate data. There are a number of methods for accurately measuring temperature, as given in reference 65(a), pp. 71-76 or reference 5(b), pp. 331-339.

Spin Saturation Transfer

The technique of spin saturation transfer (or magnetization transfer) was first developed by Forsén and Hoffman [68]. Many very detailed reviews on the subject are available [65], including some useful experimental tips [5(b), 65(a)]. This method allows the measurement of rates down to 10^{-1}s^{-1}, an order of magnitude slower than those obtainable from line shape analysis.

Consider a simple two-site exchange with equal populations and no coupling to other nuclei.

$$A \underset{k_{-1}}{\overset{k_1}{\rightleftharpoons}} B$$

Experimentally, SST involves irradiating (saturating) the signal arising from nucleus B and observing a decrease in the signal arising from nucleus **A**. Because irradiation of a nucleus equalizes the populations of the two spin states, i.e., that nucleus is "saturated", magnetization of the irradiated nucleus (**B**) goes to zero. If nucleus **B** is exchanging with nucleus **A**, some of the decrease in magnetization will be transferred to **A** if the longitudinal relaxation time (T_1) of

nucleus **A** is of the same order of magnitude or slower than transfer of the saturation from **B** into that site. If nucleus **A** relaxes back to an equilibrium spin state faster than the magnetization can be transferred to **A** via the dynamic process, no change in the signal intensity of **A** is observed. If magnetization transfer occurs much faster than relaxation of the **A** nucleus, **A** will become rapidly saturated and no signal will be observed for nucleus **A**. SST is thus applicable to a system if the lifetimes of the exchanging nuclei in a particular site are on the same order of magnitude as the spin-lattice relaxation time (T_1) of that nucleus.

Mathematically, rate constants can be extracted from the experimental data by consideration of the Bloch equation that applies to the system.

$$\frac{dM_{zA}}{dt} = \frac{M_{0A} - M_{zA}}{T_{1A}} - \frac{M_{zA}}{\tau_A} + \frac{M_{zB}}{\tau_B} \qquad (10)$$

where M_{0A} = the equilibrium magnetization of the nuclei at site **A** before perturbation by radio frequency energy

M_{zA} = magnetization of nucleus **A** at time t after perturbation by radio frequency energy

M_{zB} = magnetization of nucleus **B** at time t after perturbation by radio frequency energy

T_{1A} = longitudinal relaxation time of nucleus **A**

τ_A = the lifetime of nucleus A at site A

τ_B = the lifetime of nucleus B at site B

Irradiation of **B** causes the last term in equation (10) to go to zero, and the resulting equation (11) gives an expression for the change in magnetization of nucleus **A** with time.

$$\frac{dM_{zA}}{dt} = \frac{M_{0A}}{T_{1A}} - \frac{M_{zA}}{\tau_{1A}} \qquad (11)$$

where τ_{1A} is defined by equation (12).

$$\frac{1}{\tau_{1A}} = \frac{1}{T_{1A}} + \frac{1}{\tau_A} \qquad (12)$$

Before perturbation of the system, $M_{zA} = M_{0A}$ and $t=0$, and under these conditions equation (11) integrates to

$$M_{zA} = M_{0A} \left[\frac{\tau_{1A}}{T_{1A}} + \frac{\tau_{1A}}{\tau_A} \exp(-t/\tau_{1A}) \right] \tag{13}$$

When the equilibrium magnetization is reached, the equation becomes

$$M_{zA} = M_{0A\infty} \left(\frac{\tau_{1A}}{T_{1A}} \right) \tag{14}$$

There are two methods for obtaining rate constants from these experiments. The first method involves measuring the decrease in magnetization of A (M_{zA}) as this value approaches the equilibrium magnetization, $M_{zA\infty}$. A semilogarithmic plot of equation (13) as $\ln(M_{zA} - M_{zA\infty})$ versus time gives a straight line, with the slope of the line being $-1/\tau_{1A}$. Since the area under the peak is proportional to the magnetization, the value of M_{zA} is simply the integral of the peak for nucleus A after irradiating nucleus B for time t. The value of $M_{zA\infty}$ is the integral of the peak for nucleus A after equilibrium magnetization has been reached. The value of τ_{1A} obtained from the plot and the value of T_{1A} (obtained from a separate T_1 measurement (see references 65(a) and 5(b)) can be substituted into equation (12) to obtain a value for τ_A, which is the inverse of the rate constant, k. Alternatively, nucleus B can be irradiated for long enough to ensure that the magnetization of A is at the equilibrium magnetization, $M_{zA\infty}$, and equations (14) and (12) can be combined to obtain the expression for the rate constant given in equation (15):

$$k = \frac{1}{T_{1A}} \left[\frac{M_{0A}}{M_{zA\infty}} - 1 \right] \tag{15}$$

Some applications of SST to organometallic systems are given in reference 69, and Chapter 4 of this book provides some practice problems for both line shape analysis and spin saturation transfer.

Calculating Activation Parameters

The rate constants obtained by the methods above can be used to calculate activation parameters for the fluxional process being studied. The Eyring equation (16) gives the relationship of the rate constant to the free energy of activation, $\Delta G^{\ddagger}$.

$$k = (\kappa T/h) \, e^{-\Delta G^{\ddagger}/RT} \tag{16}$$

thus, $\Delta G^{\ddagger} = -RT[\ln(k/T) + \ln(h/\kappa)]$ (17)

where $R = 1.987 \times 10^{-3}$ kcal/mol K

κ = Boltzmann's constant = 1.38054×10^{-16} erg/K

h = Plank's constant = 6.6256×10^{-27} erg sec

T = temperature in K

The activation enthalpy $\Delta H^{\ddagger}$ and activation entropy $\Delta S^{\ddagger}$ can be calculated from the temperature dependence of the rate constant according to equation (18).

$$k = (\kappa T/h)\, e^{-\Delta H^{\ddagger}/RT}\, e^{\Delta S^{\ddagger}/R}$$ (18)

The numbers for $\Delta G^{\ddagger}$ obtained in this manner are fairly accurate; indeed, errors of $5^{o}C$ in temperature and 15% in k will still give reasonably accurate values of $\Delta G^{\ddagger}$; however, in order to obtain accurate $\Delta H^{\ddagger}$ and $\Delta S^{\ddagger}$ values, it is much more important to have very precise k versus T data over a wide temperature range.

MISCELLANEOUS ITEMS OF INTEREST

There are a number of NMR techniques that are beyond the scope of this book but may be applicable to certain systems. We thought it would be useful to have a leading reference to some of these techniques.

1) Solid state NMR can be applied to the study of polymer-bound organometallics and heterogeneous catalysts [70].

2) NMR spectra of paramagnetic molecules typically show broad lines that are spread out over a wide range (> 200 ppm for protons), or sharp lines that are shifted out of the usual range being observed. Paramagnetic NMR spectroscopy can give information about species that are difficult to characterize otherwise; however, special parameters for acquisition and processing of the spectrum are required [71].

3) Detection of radical pairs as intermediates in a reaction is often possible by the use of Chemically Induced Dynamic Nuclear Polarization (CIDNP) [72].

4) A wide variety of new pulse techniques are now available for obtaining information about couplings and exchange processes, including SEFT (spin echo fourier transform), SPI (selective population inversion), INEPT (insensitive nuclei enhanced by polarization transfer), DEPT (distortionless enhancement by polarization transfer), INADEQUATE (incredible natural abundance double quantum transfer experiment), and DANTE (delays alternating with nutation for tailored excitation). This information has been summarized through 1983 in reference 73, as well as descriptions of useful 2-D NMR techniques such as COSY (correlated spectroscopy), SECSY (spin echo correlated spectroscopy), 2-D INADEQUATE, and NOESY (nuclear Overhauser effect spectroscopy).

REFERENCES

1. (a) Wilkinson, G.; Stone, F.G.A; Abel, E.W., eds. "Comprehensive Organometallic Chemistry", v. 1-9, Pergamon Press, (Oxford, 1982).

(b) Bau, R.; Teller, R.G.; Kirtley, S.W.; Koetzle, T.F. Accts. Chem. Res. **1979**, 12, 176.

2. (a) Connelly, N.G.; Geiger, W. Adv. Organomet. Chem. **1985**, 24, 87.

(b) Connelly, N.G.; Geiger, W. Adv. Organomet. Chem. **1984**, 23, 1.

(c) Litzow, M.R.; Spaulding, T.R. "Mass Spectroscopy of Inorganic and Organometallic Compounds", Elsevier (Amsterdam, 1973).

(d) Sharp, T.R.; White, M.R.; Davis, J.F.; Stang, P.J. Org. Mass Spec. **1984**, 19, 107.

(e) Lay, J.O., Jr.; Allison, N.T.; Yongskulrote, W.; Ferede, R. Org. Mass Spec. **1986**, 21, 371.

(f) Rapic, V.; Filipović-Marinić, N. Org. Mass Spec. **1985**, 20, 688.

3. (a) Abragam, A., "The Principles of Nuclear Magnetism", Oxford University Press (Oxford, 1961).

(b) Slichter, C.P., "Principles of Magnetic Resonance", Springer-Verlag (Berlin, 1978).

(c) Farrar, T.C.; Becker, E.D. "Pulse and Fourier Transform NMR", Academic Press (New York, 1971).

(d) Ando, I.; Webb, G.A. "Theory of NMR Parameters", Academic Press (New York, 1983)

(e) Schumacher, R.T. "Introduction to Magnetic Resonance", W.A. Benjamin, Inc. (New York, 1970).

4. (a) Abraham, R.J.; Loftus, P. "Proton and Carbon-13 NMR Spectroscopy", John Wiley and Sons (New York, 1983).

(b) Günther, H. "NMR Spectroscopy. An Introduction", John Wiley and Sons (New York, 1980).

(c) Becker, E.D. "High Resolution NMR" Academic Press (New York, 1980).

5. (a) Fukushima, E.; Roeder, S.B.W. "Experimental Pulse NMR, A Nuts and Bolts Approach", Addison-Wesley (Reading, MA, 1981).

(b) Martin, M.L.; Delpuech, J.-J.; Martin, G.J. "Practical NMR Spectroscopy", Heyden (Philadelphia, 1980).

5. (c) Brevard, C. "NMR of Newly Accessible Nuclei", Vol. 1, P. Laszlo, ed., Academic Press (New York, 1983), p. 1.

(d) Bolton, P.H. "NMR of Newly Accessible Nuclei", Vol. 1, P. Laszlo, ed., Academic Press (New York, 1983), p.21.

6. For an extensive collection of both proton and carbon-13 NMR data for specific organometallic complexes, see Hickmott, P.W.; Cais, M.; Modiano, A. Ann. Rep. Nuc. Mag. Res. Spec. 1978, 6C.

7. Carbon-13: (a) Levy, G.C. "Carbon-13 Nuclear Magnetic Resonance for Organic Chemists", Wiley-Interscience (New York, 1972).

(b) Chisholm, M.H.; Godleski, S. Prog. Inorg. Chem. 1976, 20, 299.

(c) Mann, B.E. Adv. Organomet. Chem. 1974, 12, 135.

(d) Jolly, P.W.; Mynott, R. Adv. Organomet. Chem. 1981, 19, 257.

(e) Harris, R.K.; Mann, B.E., eds. "NMR and the Periodic Table", Academic Press (New York, 1978).

(f) For a collection of Carbon-13 data for organometallic complexes, see Mann, B.E.; Taylor, B.F. "Carbon-13 NMR Data for Organometallic Compounds", Academic Press (New York, 1981).

8. Nitrogen-15: (a) Levy, G.C.; Lichter, R.L. "Nitrogen-15 Nuclear Magnetic Resonance Spectroscopy", Wiley-Interscience (New York, 1979).

(b) Martin, G.J.; Martin, M.L.; Gouesnard, J.-P. NMR, Basic Principles and Progress, v. 18, Springer-Verlag (Berlin, 1981).

(c) Brevard, C.; Granger, P. "Handbook of High Resolution NMR", John Wiley and Sons (New York, 1981).

9. Fluorine-19: (a) Fields, R. Ann. Rep. Nuc. Mag. Res. Spec. 1977, 7, 1.

(b) Cavalli, L. Ann. Rep. Nuc. Mag. Res. Spec. 1976, 6B, 43.

(c) Wray, V. Ann. Rep. Nuc. Mag. Res. Spec. 1980, 10B, 1.

10. Silicon-29: (a) Williams, E.A.; Cargioli, J.D. Ann. Rep. Nuc. Mag. Res. Spec. 1979, 9, 221.

(b) Coleman, B. "NMR of Newly Accessible Nuclei", Vol. 2, P. Laszlo, ed., Academic Press (New York, 1983), p. 197.

(c) Thomas, J.M.; Klinowski, J. Adv. in Catalysis 1985, 33, 199.

(d) Blinka, T.A.; Helmer, B.J.; West, R. Adv. Organomet. Chem. 1984, 23, 193.

10. (e) Williams, E.A. <u>Ann</u>. <u>Rep</u>. <u>Nuc</u>. <u>Mag</u>. <u>Res</u>. <u>Spec</u>. **1983**, <u>15</u>, 235.

(f) Marsmann, H. <u>NMR</u>, <u>Basic</u> <u>Principles</u> <u>and</u> <u>Progress</u>, v. 17, Springer-Verlag (Berlin, 1981), 65.

(g) See also references 7(e) and 8(c).

11. **Phosphorus-31:** (a) Pregosin, P.E.; Kunz, R.W. "^{31}P and ^{13}C NMR of Transistion Metal Phosphine Complexes", Springer-Verlag (Berlin, 1979).

(b) Garrou, P.E. <u>Chem</u>. <u>Rev</u>. **1981**, <u>81</u>, 229.

(c) Derencsenyi, T.T. <u>Inorg</u>. <u>Chem</u>. **1981**, <u>20</u>, 665.

(d) Meek, D.W.; Mazanec, T.J. <u>Accts</u>. <u>Chem</u>. <u>Res</u>. **1981**, <u>14</u>, 266.

(e) See also references 7(e) and 8(c).

12. **Iron-57:** (a) von Philipsborn, W. <u>Pure</u> <u>and</u> <u>Appl</u>. <u>Chem</u>. **1986**, <u>58</u>, 513.

(b) Granger, P. "NMR of Newly Accessible Nuclei", Vol. 2, P. Laszlo, ed., Academic Press (New York, 1983), p. 386.

(c) See also references 7(e) and 8(c).

13. **Selenium-77, Tellurium-123, Tellurium-125:** (a) McFarlane, H.C.E.; McFarlane, W. "NMR of Newly Accessible Nuclei", Vol. 2, P. Laszlo, ed., Academic Press (New York, 1983), p. 275.

(b) See also references 7(e) and 8(c).

14. **Yttrium-89:** (a) Kidd, R.G. <u>Ann</u>. <u>Rep</u>. <u>Nuc</u>. <u>Mag</u>. <u>Res</u>. <u>Spec</u>. **1980**, <u>10A</u>, 1.

(b) See also references 7(e) and 8(c).

15. **Rhodium-103:** (a) Mann, B.E. "NMR of Newly Accessible Nuclei", Vol. 2, P. Laszlo, ed., Academic Press (New York, 1983), p. 301.

(b) Von Philipsborn, W. <u>Pure</u> <u>and</u> <u>Appl</u>. <u>Chem</u>. **1986**, <u>58</u>, 513.

(c) See also references 7(e) and 8(c).

16. **Silver-107, Silver-109:** (a) Henrichs, P.M. "NMR of Newly Accessible Nuclei", Vol. 2, P. Laszlo, ed., Academic Press (New York, 1983), p. 299.

(b) See also references 7(e) and 8(c).

17. **Cadmium-111, Cadmium-113:** (a) Armitage, I.M.; Boulanger, Y. "NMR of Newly Accessible Nuclei", Vol. 2, P. Laszlo, ed., Academic Press (New York, 1983), p. 337.

17. (b) See also references 7(e) and 8(c).

18. **Tin—117, Tin—119:** (a) Smith, P.J.; Tupčiauskas, A.P. <u>Ann.</u> <u>Rep.</u> <u>Nuc.</u> <u>Mag.</u> <u>Res.</u> <u>Spec.</u> **1978**, <u>8</u>, 292.

(b) Dechter, J.J. <u>Prog.</u> <u>Inorg.</u> <u>Chem.</u> **1982**, <u>29</u>, 285.

19. **Tungsten—183, Molybdenum—95, Molybdenum—97, Chromium—53:** (a) Minelli, M.; Enemark, J.H.; Brownlee, R.T.C.; O'Connor, M.J.; Wedd, A.G. <u>Coord.</u> <u>Chem.</u> <u>Revs.</u> **1985**, <u>68</u>, 169.

(b) Faller, J.W.; Whitmore, B.C. <u>Organomet.</u> **1986**, <u>5</u>, 752.

(c) See also references 7(e), 8(c), 12(b) and 14(a).

20. **Platinum—195, Mercury—199, Scandium—45, Titanium—47, Titanium—49, Vanadium—51, Manganese—55, Zinc—67, Niobium—93, Lanthanum—139, Tantalum—181, Rhenium—185, Rhenium—187:** See references 7(e), 8(c), and 14(a).

21. **Thallium—203, Thallium—205:** (a) Hinton, J.F. Metz, K.R. "NMR of Newly Accessible Nuclei", Vol. **2**, P. Laszlo, ed., Academic Press (New York, 1983), p. 367.

(b) Hinton, J.F.; Metz, K.R., Briggs, R.W. <u>Ann.</u> <u>Rep.</u> <u>Nuc.</u> <u>Mag.</u> <u>Res.</u> <u>Spec.</u> **1982**, <u>13A</u>.

(c) See also references 18(b), 7(e), and 8(c).

22. **Lead—207, Sulfur—33, Ruthenium—99, Ruthenium—101:** See references 7(e), 8(c), and 12(b).

23. **Deuterium:** (a) Smith, I.C.P. "NMR of Newly Accessible Nuclei", Vol. **2**, P. Laszlo, ed., Academic Press (New York, 1983), p. 1.

(b) Smith, I.C.P.; Mantsch, H.H. "NMR Spectroscopy: New Methods and Applications", G.C. Levy, ed., <u>ACS</u> <u>Symp.</u> <u>Ser.</u> **1982**, <u>191</u>, p. 97.

(c) See also references 7(e) and 8(c).

24. **Boron—10 and Boron—11:** (a) Kidd, R.G. "NMR of Newly Accessible Nuclei", Vol. **2**, P. Laszlo, ed., Academic Press (New York, 1983), p. 49.

(b) Nöth, H; Wrackmeyer, B. <u>NMR, Basic Principles and Progress</u>, Vol. <u>14</u>, Springer-Verlag (Berlin, 1978), 1.

(c) See also references 7(e) and 8(c).

25. **Nitrogen—14:** See references 7(e) and 8(c).

26. **Oxygen—17:** (a) Kintzinger, J.-P. "NMR of Newly Accessible Nuclei", Vol. **2**, P. Laszlo, ed., Academic Press (New York, 1983), p. 79.

26. (b) Kintzinger, J.-P. NMR, Basic Principles and Progress, Vol. 17, Springer-Verlag (Berlin, 1981), 1.

(c) See also references 7(e) and 8(c).

27. **Aluminium-27:** (a) Delpuech, J.J. "NMR of Newly Accessible Nuclei", Vol. 2, P. Laszlo, ed., Academic Press (New York, 1983), p. 153.

(b) See also references 7(e), 8(c), and 10(c).

28. **Cobalt-59:** (a) Laszlo, P. "NMR of Newly Accessible Nuclei", Vol. 2, P. Laszlo, ed., Academic Press (New York, 1983), p. 254.

(b) Benn, R.; Cibura, K.; Hofmann, P.; Jonas, K; Rufinska, A. Organomet. **1985,** 4, 2214.

(c) See also references 7(e), 8(c), 14(a), and 15(c).

29. **Nickel-61, Gallium-69, Gallium-71, Germanium-73, Arsenic-75, Zirconium-91, Technetium-99, Palladium-105, Indium-113, Antimony-121, Antimony-123, Hafnium-177, Hafnium-179, Osmium-189, Iridium-191, Iridium-193, Gold-197, Bismuth-209:** See references 7(e) and 8(c).

30. **Copper-63, Copper-65:** See references 7(e), 8(c), 12(b), and 14(a).

31. Cotton, F.A.; Hunter, D.L.; White, A.J. Inorg. Chem. **1975,** 14, 703.

32. Todd, L.J.; Wilkinson, J.R. J. Organomet. Chem. **1974,** 80, C31.

33. (a) Bax, A.; Lerner, L. Science **1986,** 232, 960.

(b) Bax, A. "Two-Dimensional Nuclear Magnetic Resonance in Liquids" D. Reidel Publishing Co. (Boston, 1982).

(c) Ernst, R.R. ACS Symp. Ser. **1982,** 191, 47.

34. (a) Top, S.; Jaouen, G.; Vessieres, A.; Abjean, J.-P.; Davoust, D.; Rodger, C.A.; Sayer, B.G.; McGlinchey, M.J. Organomet. **1985,** 4, 2143.

(b) Cabeza, J.A.; Mann, B.E.; Brevard, C.; Maitlis, P.M. J. Chem. Soc., Chem. Comm. **1985,** 65.

(c) Kook, A.M.; Nickias, P.N.; Selegue, J.P.; Smith, S.L. Organomet. **1984,** 3, 499.

(d) Erker, G.; Mühlenbernd, T.; Benn, R.; Rufinska, A. Organomet. **1986,** 5, 402.

35. (a) Gordon, A.J.; Ford, R.A. "The Chemist's Companion", John Wiley and Sons (New York, 1972).

(b) See reference 43.

36. "The Sadtler Handbook of Proton NMR Spectra", W.W. Simons, ed., Sadtler Research Laboratories (Philadelphia, 1978).

37. Sergeyev, N.M. Prog. Nuc. Mag. Res. Spec. 1975, 9, 71.

38. (a) Manriquez, J.M.; Fagan, P.J.; Schertz, L.D.; Marks, T. Inorg. Syn. 1982, 21, 181.

(b) in THF-d$_8$, Bercaw, J.E., Burger, B., private communication.

39. (a) Vogel, E.; Altenbach, H.-J.; Drossard, J.-M.; Schmickler, H.; Stegelmeier, H. Angew. Chem. Int. Ed. Engl. 1980, 19, 1016.

(b) Wehner, R.; Günther, H. Chem. Ber. 1974, 107, 3152.

40. (a) Takeuchi, K.; Yokomichi, Y; Kubota, Y.; Okamoto, K. Tetrahedron, 1980, 36, 2939.

(b) Cox, R.H.; Harrison, L.W.; Austin, W.K., Jr. J. Phys. Chem. 1973, 77, 200.

41. (a) Baker, R.T.; Tulip, T.H. Organomet. 1986, 5, 839.

(b) Kohler, F.H. Chem. Ber. 1974, 107, 570.

42. (a) Templeton, J.L.; Herrick, R.S.; Morrow, J.R. Organomet. 1984, 3, 535.

(b) Templeton,J.L.; Ward, B.C. J. Am. Chem. Soc. 1980, 102, 3288.

(c) Herrick, R.S.; Templeton, J.L. Organomet. 1982, 1, 842.

43. Bailey, W.I.,Jr.; Chisholm, M.H.; Cotton, F.A.; Rankel, L.A. J. Am. Chem. Soc. 1978, 100, 5764.

44. (a) Thomas, J.L. Inorg. Chem. 1978, 17, 1507.

(b) Brown, L.D.; Barnard, C.F.J.; Daniels, J.A.; Mawby, R.J.; Ibers, J.A. Inorg. Chem. 1978, 17, 2932.

(c) Schrock, R.R.; Sharp, P.R. J. Am. Chem. Soc. 1978, 100, 2389.

(d) Guggenberger, L.J ; Meakin, P.; Tebbe, F.N. J. Am. Chem. Soc. 1974, 96, 5420.

(e) Faller, J.W.; Johnson, B.V. J. Organomet. Chem. 1975, 88, 101.

45. (a) Eschbach, C.S.; Seyferth, D.; Reeves, P.C. J. Organomet. Chem. 1976, 104, 363.

(b) Ville, G.A.; Vollhardt, K.P.C.; Winter, M.J. Organomet. 1984, 3, 1177.

46. (a) Isaacs, E.E.; Graham, W.A.G. J. Organomet. Chem. **1975**, **90**, 319.

(b) Groenenboom, C.J.; Jellinek, F. J. Organomet. Chem. **1974**, **80**, 229.

47. Collman, J.P.; Hegedus, L.S. "Principles and Applications of Organotransition Metal Chemistry", 1st ed., University Science Books (Mill Valley, CA, 1980).

48. Maciel, G.E.; Dallas, J.L.; Miller, D.P. J. Am. Chem. Soc. **1976**, **98**, 5074.

49. Butler, I.S. Accts. Chem. Res. **1977**, **10**, 359.

50. (a) Herrmann, W.A. Adv. Organomet. Chem. **1982**, **20**, 159.

(b) Strutz, H.; Schrock, R.R. Organomet. **1984**, **3**, 1600.

51. (a) Vaughn, G.D.; Strouse, C.E.; Gladysz, J.A. J. Am. Chem. Soc. **1986**, **108**, 1462.

(b) Vaughn, G.D.; Gladysz, J.A. J. Am. Chem. Soc. **1986**, **108**, 1473.

(c) Nelson, G.O. Organomet. **1983**, **2**, 1474.

(d) Thorn, D.L. Organomet. **1982**, **1**, 197.

52. Gladysz, J.A. Adv. Organomet. Chem. **1982**, **20**, 1.

53. Nakamoto, K. "Infrared Spectra of Inorganic and Coordination Compounds", 2nd ed., John Wiley and Sons (New York, 1970).

54. Eichhorn, G.L. "Inorganic Biochemistry", v. **2**, Elsevier (New York, 1973), 783.

55. Cotton, F.A.; Wilkinson, G. "Advanced Inorganic Chemistry", John Wiley and Sons (New York, 1980).

56. (a) Eisenberg, R.; Meyer, C.D. Accts. Chem. Res. **1975**, **8**, 26.

(b) Bottomley, F. Accts. Chem. Res. **1978**, **11**, 158.

(c) Hunter, A.D.; Legzdins, P. Organomet. **1986**, **5**, 1001.

57. (a) Herrmann, W.A.; Flöel, M.; Weber, C.; Hubbard, J.L.; Schäfer, A. J. Organomet. Chem. **1985**, **286**, 369.

(b) Seidler, M.D.; Bergman, R.G. Organomet. **1983**, **2**, 1897.

58. Socrates, G. "Infrared Characteristic Group Frequencies", John Wiley and Sons (New York, 1980).

59. Collman, J.P.; Halpert, J.R.; Suslick, K.S. "Metal Ion Activation of Dioxygen", T.G. Spiro, ed., John Wiley and Sons (New York, 1980), 49.

60. Collman, J.P.; Hegedus, L.S.; Norton, J.R.; Finke, R.G. "Principles and Applications of Organotransition Metal Chemistry", 2nd ed., University Science Books (Mill Valley, CA, 1986).

61. (a) Curtis, M.D.; Shiu, K.B.; Butler, W.M. J. Am. Chem. Soc. 1986, 108, 1550.

(b) Fachinetti, G.; Floriani, C. J. Chem. Soc., Dalt. Trans. 1977, 2297.

(c) Erker, G.; Rosenfeldt, F. Angew. Chem. Int. Ed. Engl. 1978, 17, 605.

62. Crabtree, R.H., private communication.

63. (a) Crabtree, R.H.; Lavin, M. Bonneviot, L. J. Am. Chem. Soc. 1986, 108, 4032.

(b) Crabtree, R.H.; Hamilton, D.G. J. Am. Chem. Soc. 1986, 108, 3124.

(c) Crabtree, R.H., Lavin, M. J. Chem. Soc., Chem. Comm. 1985, 1661.

(d) Crabtree, R.H.; Lavin, M. J. Chem. Soc., Chem. Comm. 1985, 794.

(e) Kubas, G.J.; Ryan, R.R.; Swanson, B.I.; Vergamini, P.J.; Wasserman, H.J. J. Am. Chem. Soc. 1984, 106, 451.

(f) Kubas, G.J.; Ryan, R.R.; Wrobleski, D.A. J. Am. Chem. Soc. 1986, 108, 1339.

(g) Morris, R.H.; Sawyer, J.F.; Shiralian, M.; Zubkowski, J.D. J. Am. Chem. Soc. 1985, 107, 5581.

64. Brookhart, M.; Green, M.L.H. J. Organomet. Chem. 1983, 250, 395.

65. (a) Sandstrom, J. "Dynamic NMR Spectroscopy", Academic Press (New York, 1982).

(b) Kaplan, J.I.; Fraenkel, G. "NMR of Chemically Exchanging Systems", Academic Press (New York, 1980).

(c) Steigel, A. NMR, Basic Principles and Progress Vol. 15, Springer-Verlag (Berlin, 1978), 1.

(d) "Dynamic Nuclear Magnetic Resonance Spectroscopy", Jackman, L.M.; Cotton, F.A., eds., Academic Press (New York, 1975).

(e) Brookhart, M., "Dynamic Processes as They Affect NMR Spectra", Short Course, University of North Carolina at Chapel Hill, April, 1980.

65. (f) Vrieze, K; Van Leeuwen, P.W.N.M. Prog. in Inorg. Chem., 1971, 14, 1.

66. Anet, F.A.L.; Bourn, A.J.R. J. Am. Chem. Soc. 1967, 89, 760.

67. Adams, R.D.; Cotton, F.A. Inorg. Chim. Acta, 1973, 7, 153.

68. (a) Forsén, S.; Hoffman, R.A. Acta Chem. Scand. 1963, 17, 1787.

(b) Forsén, S.; Hoffman, R.A. J. Chem. Phys. 1963, 39, 2892.

(c) Forsén, S.; Hoffman, R.A. J. Chem. Phys. 1964, 40, 1189.

(d) Hoffman, R.A.; Forsén, S. Prog. NMR Spec. 1966, 1, 173.

69. (a) Faller, J.W. "Determination of Organic Structures by Physical Methods", Nachod, F.C.; Zuckerman, J.J., eds., v. 5, Academic Press (New York, 1973), 75.

(b) Holmgren, J.S.; Shapley, J.R. Organomet. 1985, 4, 793.

70. Fyfe, C.A. "Solid State NMR for Chemists", C.F.C. Press, (P.O Box 1720, Guelph, Ontario, Canada, 1983).

71. (a) "NMR of Paramagnetic Molecules", LaMar, G.N.; Horrocks, W.D.; Holm, R.H., eds., Academic Press (New York, 1973).

(b) Orrell, K.G. Ann. Rep. Nuc. Mag. Res. Spec. 1979, 9, 1.

72. (a) Closs, G.L.; Miller, R.J.; Redwine, O.D. Accts. Chem. Res. 1985, 18, 196.

(b) "Chemically Induced Magnetic Polarization", Lepley, A.R.; Closs, G.L., eds., John Wiley and Sons (New York, 1973).

73. Benn, R.; Günther, H. Angew. Chem. Int. Ed. Engl. 1983, 22, 350.

2
Structure and Bonding

 2a$_1$

  2e

ML$_3$

  1e

 1a$_1$

QUESTIONS

1. Give the formal oxidation state and d-electron count of the metal in the following complexes.

a) $(\eta^6\text{-}C_6H_6)_2Mo$

b) $Cp_2ZrCl(OMe)$

c) $[(PMe_3)_2Pd(\eta^3\text{-}C_3H_3)]^+$

d) $(CO)_5Re\text{-}Et$

e) $(diphos)Pt(OMe)_2$

f) $Cp^*(PMe_3)_2Ru\text{-}Cl$

g) $(dtc)_2W(C_2H_4)(Ph\text{-}C\equiv C\text{-}Ph)$ (dtc = dithiocarbamate, see question **7.**)

h) $[Cp(PMe_3)_3WH_2]^+$

2. For each of the following species, indicate the number of electrons in the valence shell of the metal. Which of these complexes would you expect to be stable enough to be characterized?

a) $(\eta^6\text{-}C_6H_6)_2Fe$

b) $Ru(PPh_3)_2(CO)_2$

c) $Cp_2NbH(C_2H_4)$

d) $[HFe(CO)_4]^-$

e) $Re(CO)_5$

f) $Pt(PBu_3)_3$

g) $(RO)_3W\equiv CMe$

h) $[Cp(diphos)(CO)_2Mo]^+$

i) $(CO)_4(Br)W\equiv CPh$

j) $[Cp(CO)_2Fe(PhC\equiv CH)]^+$

k) $[PtCl_3(C_2H_4)]^-$

l) $[Cp(CO)(NO)Mo(\eta^3\text{-}C_3H_3)]^+$

m) $(\eta^6\text{-}C_6H_6)Mn(CO)_2(CH_3)$

n) $Cp(NO)_2W\text{-}H$

o) $[Ir(CO)(PPh_3)_2(Cl)(NO)]^+$

p) $Cp(CO)_2Re(CH_3)_2$

q)

r)

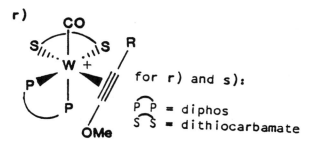

for r) and s):

$\widehat{P\ P}$ = diphos

$\widehat{S\ S}$ = dithiocarbamate

s)

3. For the following polynuclear complexes, indicate the total number of electrons, determine the number of M–M bonds present (assuming that all metals are coordinatively saturated), and predict a structure.

a) $\mu-CO-[(\eta^4-C_4H_4)Fe(CO)]_2$

b) $\mu-CO-\mu-CRR'-[Cp^*Rh]_2$

c) $\mu-CRR'-[Cp^*Rh(CO)]_2$

d) $[\mu-X-\mu-CH_2-(Os_3(CO)_{10})]^-$, where X = halide

e) $[\mu-Cl-Rh(PPh_3)(PPh_2CH_2Ph-\textcircled{P})]_2$, where $\textcircled{P}$ = polymer

f) $\mu-PPh_2[(PPh_3)(CO)_3FeIr(CO)_2(PPh_3)]$

g) $\mu^3-CO-[Fe_4(CO)_{12}]^{-2}$

h) $[\mu-Cl-PdCl(RC\equiv CR)]_2$

i) $(\mu-CO)_2-[CpRh]_3(CO)$

j) $(\mu-Br)_2-[Mn(CO)_4]_2$

k) $Cp_2(t-BuO)ZrRu(CO)_2Cp$

l) $(\mu-CH_2)_2-[Cp^*Rh(CH_3)]_2$

m) $[Cp(CO)_2Mo]_2$

n) $[\mu-H]_2-[Cp(NO)W(H)]_2$

4. For each of the following metal and ligand combinations, formulate the simplest neutral compound that conforms to the 18–electron rule. Draw a reasonable structure for each compound.

a) Ni, CO   **b)** Fe, CO, COT = cyclooctatetraene

c) Co, Cp, CO   **d)** Fe, Cp, CO

e) Re, CO, H   **f)** Mo, Cp, CO

g) Ni, Cp, NO   **h)** Cp, Co, NO

5. Predict the structures of the complexes given.

a) $Ni(CO)_4$ **b)** $(PEt_3)_2Pd(CH_3)_2$

c) $Fe(CO)_4(C_2H_4)$ **d)** $PdCl_2(CH_3CN)_2$

e) $Rh(PPh_3)_3Cl$ **f)** $(PPh_3)_2Pt(Ph-C\equiv C-Ph)$

g) $[Co(CO)_4]^-$ **h)** $[Ir(diphos)_2]^+$

5. (cont.)

i) $(Ph)_2Ni(PPh_3)_2$ **j)** $(dmpe)_2Fe(PPh_3)$

6. Although cobaltocene (Cp_2Co) is stable (under N_2) at room temperature, the rhodium and iridium analogs dimerize to give the compounds $M_2C_{20}H_{20}$. Three plausible structures are shown below. Are these 18-electron species? Use the following NMR data to determine the correct structure.

^{1}H NMR: 5.2 (s, 10 H)
              5.0 (m, 4 H)
              3.3 (m, 4 H)
              2.2 (m, 2 H)

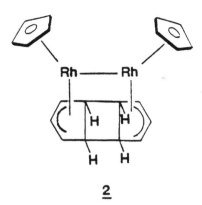

2

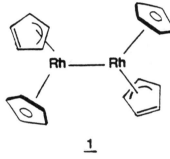

1

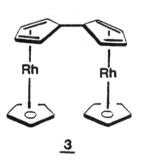

3

7. Determine the number of valence shell electrons in the metal-acetylene complexes below and indicate how many electrons the acetylene will donate to the metal.

a) $W(CO)(detc)_2(HC\equiv CH)$ (detc = diethyldithiocarbamate, illustrated below)

b) $CpW(CO)(CH_3)(HC\equiv CH)$

c) $Cp_2Mo(HC\equiv CH)$

d) $Mo(detc)_2(HC\equiv CPh)_2$

detc

8. For each pair of complexes, explain the relative values of $\Delta G^{\ddagger}$ for rotation around the metal—carbon bond. In all cases, there is only one d-orbital hybrid available for pi interaction with the carbene ligand.

a) $[Cp(CO)_2(PPh_3)Mo=CH_2]^+$, $\Delta G^{\ddagger}_{rot} < 6$ kcal/mol
vs
$[Cp(CO)_2(PPh_3)W=CH_2]^+$, $\Delta G^{\ddagger}_{rot} = 8.3$ kcal/mol

b) $[Cp(CO)_2(PPh_3)W=CH_2]^+$, $\Delta G^{\ddagger}_{rot} = 8.3$ kcal/mol
vs
$[Cp(CO)_2(PEt_3)W=CH_2]^+$, $\Delta G^{\ddagger}_{rot} = 9.0$ kcal/mol

c) $[Cp(diphos)Fe=CH_2]^+$, $\Delta G^{\ddagger}_{rot} = 10.4$ kcal/mol
vs
$[Cp(NO)(PPh_3)Re=CH_2]^+$, $\Delta G^{\ddagger}_{rot} > 15$ kcal/mol

9. Which of the two complexes has the lower energy CO stretching frequency in the infrared spectrum? Rationalize your choice.

a) $[W(CO)_5Cl]^-$ or $Re(CO)_5Cl$

b) $Fe(CO)_5$ or $Fe(CO)_4Br_2$

c) $Mo(CO)_6$ or $Mo(CO)_4(PPh_3)_2$

d) $Mo(CO)_4(PMe_3)_2$ or $Mo(CO)_4(PPh_3)_2$

e) $Cp(CO)_2Fe-Br$ or $[Cp(CO)_2Fe]^-$

10. Using the Dewar—Chatt—Duncanson model of bonding, predict the orientation of the ethylene in each of the following complexes. The metal fragment orbitals and the pi orbitals of ethylene are shown on the following page. Explain your reasoning.

a) $(ethylene)Ni(PPh_3)_2$

b) $(ethylene)Fe(CO)_4$

c) $[(ethylene)PtCl_3]^-$

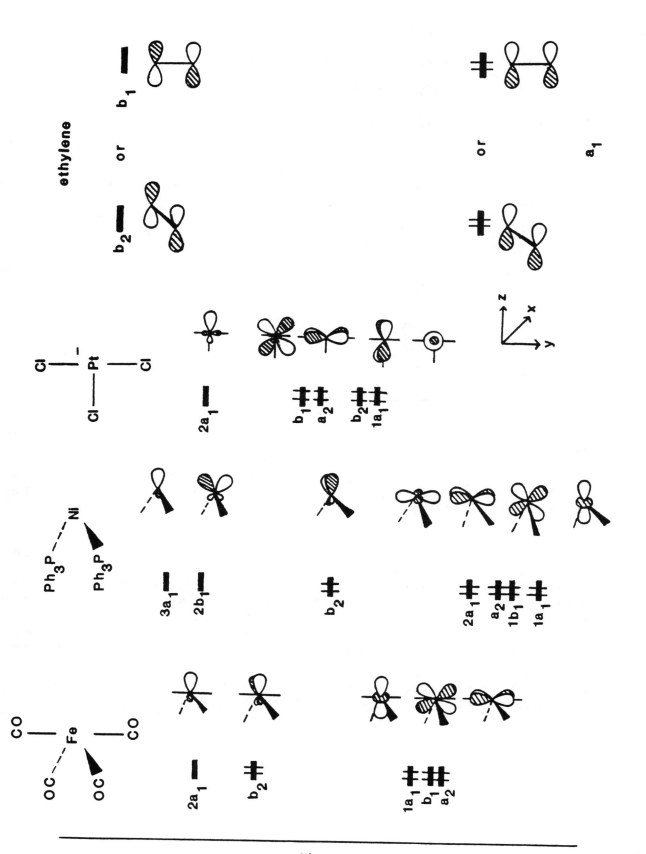

11. Equilibrium constants for the reaction

$$CoBr_2(CO)L_2 \rightleftharpoons CoBr_2L_2 + CO$$

for various L are given below.

L	K_d	$CO(cm^{-1})$	cone angle (degrees)
PEt_3	(1)	1985	132
$P(n-Pr)_3$	1.1	1980	135
PEt_2Ph	2.5	1990	135
$PEtPh_2$	24.2	1990	140
PPh_3	750.0	1995	145

a) Is $CoBr_2(CO)L_2$ an 18-electron complex?

b) Rationalize the trends observed for K_d.

12. Choose the appropriate molecular orbital diagram for the molecular orbitals of the following metal fragments and find the hydrocarbon isolobal analog for each one. Molecular orbital diagrams for the possible hydrocarbon fragments and the various metal fragments are given on the following pages.

HINTS
1) The isolobal analogy allows one to compare systems that contain the same number of orbitals and electrons.
2) In using this analogy, it is necessary to decide which orbitals are important for bonding to an additional ligand or substituent i.e., the orbitals already involved in bond formation to other groups can be ignored, as well as any antibonding orbitals. The nonbonding orbitals of the metal fragment (those derived from the t_{2g} set) can also be ignored, since they are only involved in pi bonding and are therefore not available for sigma-type interactions with an incoming ligand.
3) If you are still stuck, see the answer to **a)** for an example.

a) $[PtCl_3]^-$

b) $Fe(CO)_3$

c) $Fe(CO)_4$

d) $Cp(CO)Rh$

12. (cont.)

e) $[PdL_2]^{+2}$

f) $[CpFe]^+$

g) $Os(CO)_4$

h) $Re(CO)_5$

i) $Ni(CO)_2$

Hydrocarbon Fragment Molecular Orbital Diagrams

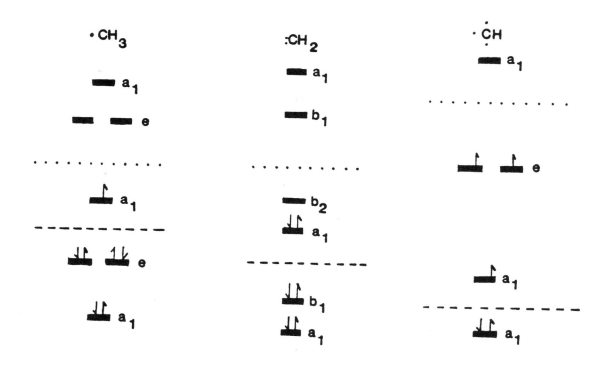

12. (cont.)

A. ML$_5$ (Octahedral minus
    one ligand)

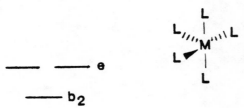

—— a$_1$

—— —— e

—— b$_2$

B. ML$_4$ (Octahedral minus
    two <u>cis</u> ligands)

—— 2a$_1$

—— 1b$_2$

—— 1a$_1$
—— 1b$_1$
—— a$_2$

C. ML$_3$ (Octahedral minus
    three facial ligands)

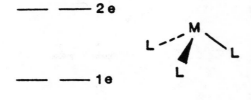

—— 2a$_1$

—— —— 2e

—— —— 1e

—— 1a$_1$

D. ML$_3$ (ML$_5$ minus two
    <u>trans</u> ligands)

—— 2a$_1$

—— 1b$_2$
—— 1a$_2$

—— 1b$_1$
—— 1a$_1$

E. ML$_2$ (ML$_4$ minus two
    axial ligands)

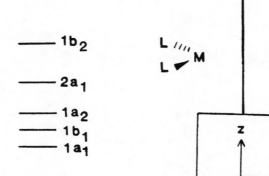

—— 3a$_1$

—— 1b$_2$

—— 2a$_1$

—— 1a$_2$
—— 1b$_1$
—— 1a$_1$

13. Using the orbitals of an ML_3 fragment and the pi molecular orbitals of the cyclopropenyl cation given below, derive the molecular orbitals of $[(cyclopropenyl)Ni(PPh_3)_3]^+$. Clearly explain:

a) which orbitals of each fragment interact to give the molecular orbitals of the final molecule

b) the proper relative orientation of the two C-3 rotors.

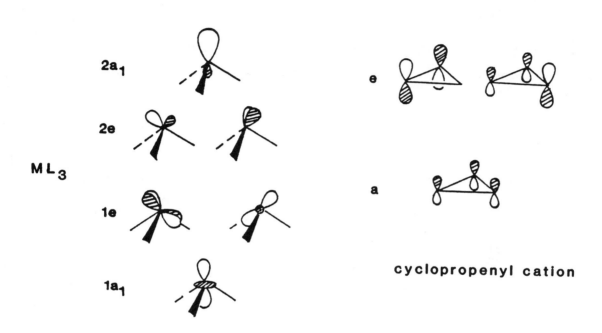

cyclopropenyl cation

14. For the formyl complex $Cp(NO)(PPh_3)Re-CHO$, there are two possible orientations of the formyl ligand, **A** or **B**. Choose the correct orientation and explain your reasoning.

HINT

In an asymmetric CpMLL' system, there are two metal d-orbital hybrids (I and II, see following page) able to participate in M-L pi interactions.

14. (cont.)

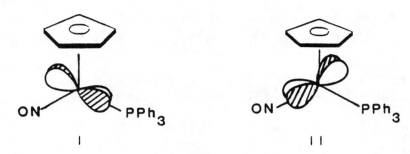

I   II

15. Using the orbitals of an ML_3 fragment (see problem 13) and the pi molecular orbitals of trimethylenemethane (TMM) shown below, construct a qualitative molecular orbital interaction diagram for $(TMM)Fe(CO)_3$.

Which is the proper relationship between the two 3-fold rotors of the molecule, **1** or **2**?

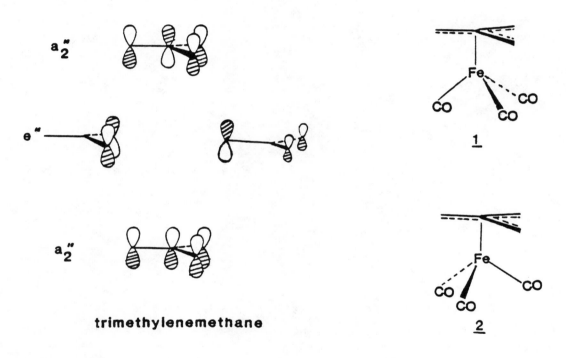

trimethylenemethane

ANSWERS

1. a) Mo(0), d^6. b) Zr(IV), d^0. c) Pd(II), d^8. d) Re(I), d^6.
e) Pt(II), d^8. f) Ru(II), d^6. g) W(II), d^4. h) W(IV), d^2.

2. a) $(\eta^6-C_6H_6)_2Fe$

2	$(\eta^6-C_6H_6)$	12 electrons	neutral
	Fe(0)	8 electrons	neutral
Total		20 electrons	

Unstable, a 20-electron complex; however, the 18- electron Fe(II) species $[(\eta^6-C_6H_6)_2Fe]^{+2}$ is well known.

b) $Ru(PPh_3)_2(CO)_2$

2	CO	4 electrons	neutral
2	PPh_3	4 electrons	neutral
	Ru(0)	8 electrons	neutral
Total		16 electrons	

Unstable, a 16-electron complex.

c) $Cp_2NbH(C_2H_4)$

2	Cp^-	12 electrons	-2
	H^-	2 electrons	-1
	(C_2H_4)	2 electrons	neutral
	Nb(III)	2 electrons	$+3$
Total		18 electrons	

Observable 18-electron complex.

2. d) $[HFe(CO)_4]^-$

4	CO	8 electrons	neutral
	H^-	2 electrons	− 1
	Fe(0)	<u>8 electrons</u>	neutral
Total		18 electrons	

Observable 18-electron complex.

e) $Re(CO)_5$

5	CO	10 electrons	neutral
	Re(0)	<u>7 electrons</u>	neutral
Total		17 electrons	

Unstable, a 17-electron species. Postulated as a radical intermediate
in reactions which involve photolytic cleavage of the Re−Re bond of
$Re_2(CO)_{10}$.

f) $Pt(PBu_3)_3$

3	PBu_3	6 electrons	neutral
	Pt(0)	<u>10 electrons</u>	neutral
Total		16 electrons	

A stable 16-electron species, not uncommon for Pt and Pd.

g) $(RO)_3W{\equiv}CMe$

3	OR^-	10 electrons	− 3
	[carbyne]$^-$	4 electrons	− 1
	W(IV)	<u>2 electrons</u>	+ 4
Total		16 electrons	

At first glance, this complex may appear to be a 12-electron species;
however, in many electron-deficient complexes that contain ligands with
strong pi donating capabilities (e.g. halogens, oxygen, sulfur,
nitrogen), the filled p-orbitals on the ligand can donate electrons
into metal d-orbitals of the appropriate symmetry and increase the
electron count around the metal. In this particular example, a
16-electron configuration is obtained by forming the triple bond to the
carbyne and three sigma bonds, as well as two pi bonds resulting from
the interaction of three filled oxygen pi orbitals with two

2. g) (cont.) metal d-orbitals. The complex actually exists as a dimer (see below), which is an 18-electron species, with the lone pairs on oxygen used to make the t-butoxide bridges. (See Chisholm, M.H.; Hoffman, D.M.; Huffmann, J.C. _Inorg._ _Chem._ **1983**, <u>22</u>, 2903 and Listemann, M.L.; Schrock, R.R. _Organomet._ **1985**, <u>4</u>, 74.)

h) [Cp(diphos)Mo(CO)$_2$]$^+$

2	CO	4 electrons	neutral
	Cp$^-$	6 electrons	− 1
	diphos	4 electrons	neutral
	Mo(II)	<u>4 electrons</u>	+ 2
Total		18 electrons	

Observable 18-electron complex.

i) (CO)$_4$(Br)W≡CPh

4	CO	8 electrons	neutral
	Br$^-$	2 electrons	− 1
	[carbyne]$^-$	4 electrons	− 1
	W(II)	<u>4 electrons</u>	+ 2
Total		18 electrons	

Observable 18-electron complex.

2. j) $[Cp(CO)_2Fe(PhC\equiv CH)]^+$

2	CO	4 electrons	neutral
	Cp^-	6 electrons	-1
	$PhC\equiv CH$	2 electrons	neutral
	Fe(II)	6 electrons	$+2$
	Total	18 electrons	

Observable 18-electron complex.

k) $[PtCl_3(C_2H_4)]^-$

3	Cl^-	6 electrons	-3
	C_2H_4	2 electrons	neutral
	Pt(II)	8 electrons	$+2$
	Total	16 electrons	

A stable 16-electron species, not uncommon for Pt and Pd.

l) $[Cp(CO)(NO)Mo(\eta^3-C_3H_3)]^+$

CO	2 electrons	neutral
NO^+	2 electrons	$+1$
Cp^-	6 electrons	-1
$[(\eta^3-C_3H_3)]^-$	4 electrons	-1
Mo(II)	4 electrons	$+2$
Total	18 electrons	

Observable 18-electron complex.

m) $(\eta^6-C_6H_6)Mn(CO)_2(CH_3)$

2	CO	4 electrons	neutral
	CH_3^-	2 electrons	-1
	$(\eta^6-C_6H_6)$	6 electrons	neutral
	Mn(I)	6 electrons	$+1$
	Total	18 electrons	

2. m) (cont.) Observable 18-electron complex.

n) $Cp(NO)_2W-H$

	Cp^-	6 electrons	$- 1$
2	NO^+	4 electrons	$+ 2$
	H^-	2 electrons	$- 1$
	$W(0)$	<u>6 electrons</u>	neutral
Total		18 electrons	

Observable 18-electron complex.

o) $[Ir(CO)(PPh_3)_2(Cl)(NO)]^+$

	CO	2 electrons	neutral
2	PPh_3	4 electrons	neutral
	Cl^-	2 electrons	$- 1$
	NO^+	2 electrons	$+ 1$
	$Ir(I)$	<u>8 electrons</u>	$+ 1$
Total		18 electrons	

Observable 18-electron complex.

p) $Cp(CO)_2Re(CH_3)_2$

	Cp^-	6 electrons	$- 1$
2	CO	4 electrons	neutral
2	CH_3	4 electrons	$- 2$
	$Re(III)$	<u>4 electrons</u>	$+ 3$
Total		18 electrons	

Observable 18-electron complex.

2. q)

2 PPh$_3$	4 electrons	neutral
2 acyl$^-$	4 electrons	− 2
Cl$^-$	2 electrons	− 1
Rh(III)	6 electrons	+ 3
Total	16 electrons	

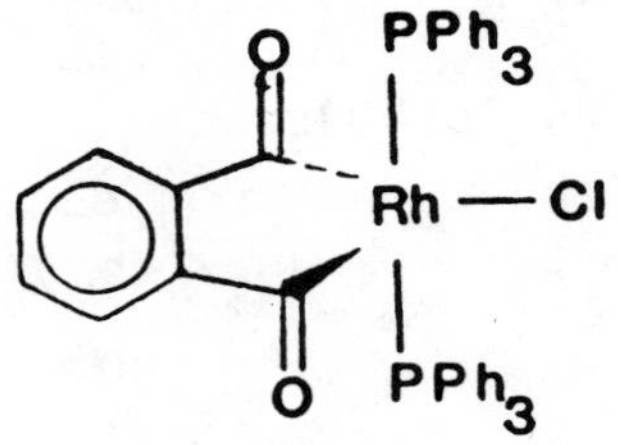

Stable 16-electron complex.

r)

dtc$^-$	4 electrons	− 1
CO	2 electrons	neutral
MeO−C≡C−R	4 electrons	neutral
diphos	4 electrons	neutral
W(II)	4 electrons	+ 2
Total	18 electrons	

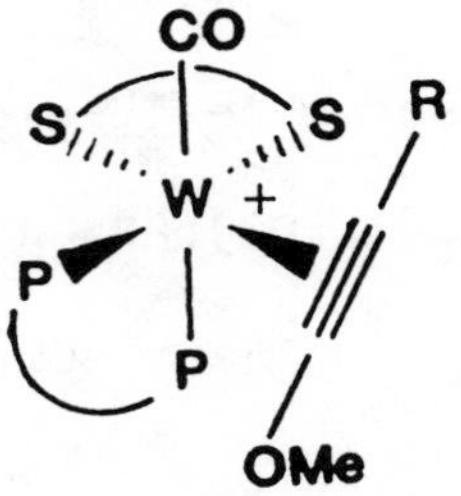

Observable 18-electron complex.

s)

dtc$^-$	4 electrons	− 1
diphos	4 electrons	neutral
CO	2 electrons	neutral
carbene	2 electrons	neutral
acyl$^-$	2 electrons	− 1
W(II)	4 electrons	+ 2
Total	18 electrons	

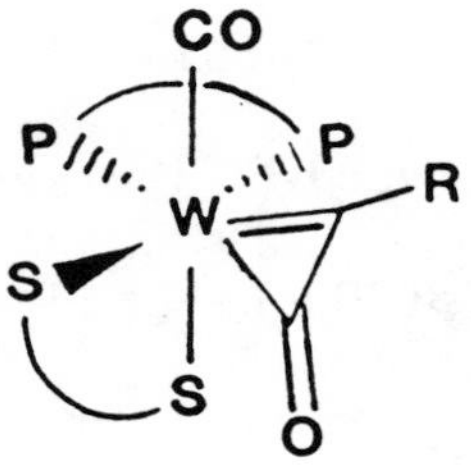

Observable 18-electron complex.

3. A way to determine the number of M—M bonds in a polynuclear complex is to determine the total number of electrons in the entire complex and subtract this number from (n x 18), where n is the number of metals in the system. The result is the number of electrons necessary to obtain an 18-electron configuration, and if it is assumed that these electrons are obtained by the formation of metal-metal bonds, division by two will give the predicted number of bonds. Not all polynuclear species fit this localized eighteen-electron bonding model. For more details on electron counting for cluster compounds, see "Transition Metal Clusters", Johnson, B.F.G., ed., Wiley-Interscience (New York, 1980).

a) μ-CO-[(η^4-C_4H_4)Fe(CO)]$_2$

	μ-CO	2 electrons	neutral
2	(η^4-C_4H_4)	8 electrons	neutral
2	CO	4 electrons	neutral
2	Fe(0)	<u>16 electrons</u>	neutral
	Total	30 electrons	

[18(2)-30]/2 = 3 Fe-Fe bonds predicted

Rest, A.J.; Sodeau, J.R.; Taylor, D.J. <u>J</u>. <u>Chem</u>. <u>Soc</u>., <u>Dalt</u>. <u>Trans</u>. **1978**, 651.

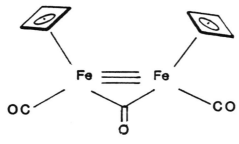

b) μ-CO-μ-CRR'-[Cp*Rh]$_2$

	μ-CO	2 electrons	neutral
	μ-CRR'	2 electrons	neutral
2	[Cp*]$^-$	12 electrons	$-$ 2
2	Rh(I)	<u>16 electrons</u>	+ 2
	Total	32 electrons	

3. b) (cont.) $[18(2)-32]/2 = 2$ Rh—Rh bonds predicted

Herrmann, W.A.; Bauer, C.; Plank, J.; Kalcher, W.; Speth, D.; Ziegler, M.L. Angew. Chem. Int. Ed. Engl. **1981**, **20**, 193.

c) μ—CRR'—[Cp*Rh(CO)]$_2$

	μ—CRR'	2 electrons	neutral
2	Cp*	12 electrons	− 2
2	CO	4 electrons	neutral
2	Rh(I)	16 electrons	+ 2
	Total	34 electrons	

$[18(2)-34]/2 = 1$ Rh—Rh bond predicted

Ibid., reference for **3b**.

d) $[\mu$—X—μ—CH$_2$—(Os$_3$(CO)$_{10}$)]$^-$

	$[\mu$—X]$^-$	4 electrons	− 1
	μ—CH$_2$	2 electrons	neutral
10	CO	20 electrons	neutral
3	Os(0)	24 electrons	neutral
	Total	50 electrons	

3. d) (cont.) [18(3)−50]/2 = 2 Os−Os bonds predicted

A halide can act as a terminal two-electron donor or a bridging four-electron donor.

Morrison, E.D.; Geoffroy, G.L.; Rheingold, A.L. <u>J</u>. <u>Am</u>. <u>Chem</u>. <u>Soc</u>. **1985**, <u>107</u>, 254.

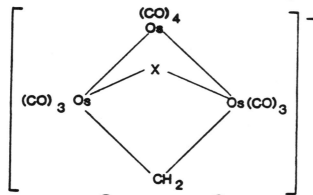

e) [μ−Cl−Rh(PPh₃)(PPh₂CH₂Ph−Ⓟ)]₂, where Ⓟ = polymer

2	[μ−Cl]⁻	8 electrons	− 2
2	PPh₃	4 electrons	neutral
2	Ph₂P(CH₂Ph−Ⓟ)	4 electrons	neutral
2	Rh(I)	<u>16 electrons</u>	+ 2
Total		32 electrons	

[18(2)−32]/2 = 2 Rh−Rh bonds predicted

Although 2 Rh−Rh bonds are predicted, the actual structure of the complex involves no M−M bonds. This is an example of a coordinatively unsaturated square planar rhodium dimer.

Collman, J.P.; Hegedus, L.S.; Cooke, M.P.; Norton, J.R.; Dolcetti, G.; Marquardt, D.N. <u>J</u>. <u>Am</u>. <u>Chem</u>. <u>Soc</u>. **1972**, <u>94</u>, 1789.

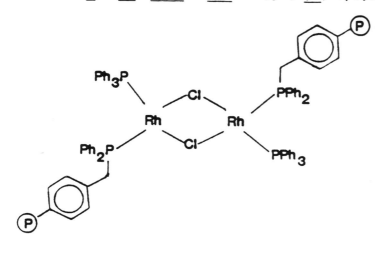

3. f) μ-PPh$_2$[(PPh$_3$)(CO)$_3$FeIr(CO)$_2$(PPh$_3$)]

2	PPh$_3$	4 electrons	neutral
5	CO	10 electrons	neutral
	[μ-PPh$_2$]$^-$	4 electrons	-1
	Fe(I)	7 electrons	$+1$
	Ir(0)	9 electrons	neutral
Total		34 electrons	

[18(2)−34]/2 = 1 Fe-Ir bond predicted

Mercer, W.C.; Whittle, R.R.; Burkhardt, E.W.; Geoffroy, G.L. <u>Organomet.</u> **1985,** <u>4</u>, 68.

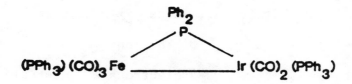

g) μ^3-CO-[Fe$_4$(CO)$_{12}$]$^{-2}$

13	CO	26 electrons	neutral
2	Fe(−1)	18 electrons	-2
2	Fe(0)	16 electrons	neutral
Total		60 electrons	

[18(4)−60]/2 = 6 Fe-Fe bonds predicted

Van Buskirk, G.; Knobler, C.B.; Kaesz, H.D. <u>Organomet.</u> **1985,** <u>4</u>, 149.

The structure was originally determined by Doedens, R.J.; Dahl, L.F. <u>J. Am. Chem. Soc.</u> **1966,** <u>88</u>, 4847.

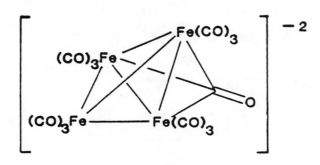

3. h) [μ–Cl–PdCl(RC≡CR)]$_2$

2	[μ–Cl]$^-$	8 electrons	– 2
2	Cl$^-$	4 electrons	– 2
2	RC≡CRR	4 electrons	neutral
2	Pd(II)	<u>16 electrons</u>	+ 4
Total		32 electrons	

[18(2)–32]/2 = 2 Pd–Pd bonds predicted

Although 2 Pd–Pd bonds are predicted based on the above calculation, the actual structure is a coordinatively unsaturated species with no metal–metal bonds. This is often the case with square-planar Pd or Pt dimers.

Collman, J.P.; Hegedus, L.S. "Principles and Applications of Organotransition Metal Chemistry", 1st ed., University Science Books (Mill Valley, CA, 1980), p. 639.

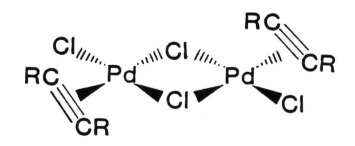

i) (μ–CO)$_2$–[CpRh]$_3$(CO)

2	μ–CO	4 electrons	neutral
3	Cp$^-$	18 electrons	– 3
	CO	2 electrons	neutral
3	Rh(I)	<u>24 electrons</u>	+ 3
Total		48 electrons	

[18(3)–48]/2 = 3 Rh–Rh bonds predicted

Lawson, R.J.; Shapley, J.R. <u>Inorg. Chem.</u> **1978**, <u>17</u>, 772.

i) (cont.)

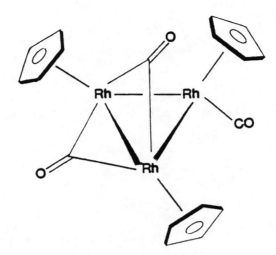

j) $(\mu-Br)_2-[Mn(CO)_4]_2$

8	CO	16 electrons	neutral
2	μ-Br	8 electrons	− 2
2	Mn(I)	12 electrons	+ 2
Total		36 electrons	

[2(18)−36]/2 = 0 Mn−Mn bonds predicted.

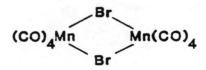

k) $Cp_2(O-t-Bu)ZrRu(CO)_2Cp$

3	Cp^-	18 electrons	− 3
	$t-BuO^-$	2 electrons	− 1
2	CO	4 electrons	neutral
	Zr(III)	1 electron	+ 3
	Ru(I)	7 electrons	+ 1
Total		32 electrons	

[2(18)−32]/2 = 2 Zr−Ru bonds predicted; however, only one is actually observed.

3. k) (cont.) Casey, C.P.; Jordan R.F. <u>J</u>. <u>Am</u>. <u>Chem</u>. <u>Soc</u>. **1983**, 105, 665.

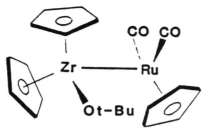

l) $(\mu\text{-}CH_2)_2\text{-}[Cp^*Rh(CH_3)]_2$

2	$\mu\text{-}CH_2^{-2}$	8 electrons	– 4
2	$[Cp^*]^-$	12 electrons	– 2
2	CH_3^-	4 electrons	– 2
2	Rh(IV)	<u>10 electrons</u>	+ 8
	Total	34 electrons	

$[2(18)-34]/2 = 1$ Rh–Rh bond predicted.

Okeya, S.; Taylor, B.F.; Maitlis, P.M. <u>J</u>. <u>Chem</u>. <u>Soc</u>., <u>Chem</u>. <u>Comm</u>. **1983**, 971.

Mann, B.E.; Meanwell, N.J.; Spencer, C.M.; Taylor, B.F.; Maitlis, P.M. <u>J</u>. <u>Chem</u>. <u>Soc</u>., <u>Dalt</u>. <u>Trans</u>. **1985**, 1555.

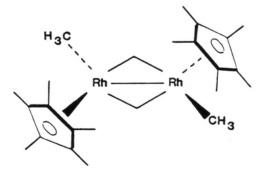

m) $[Cp(CO)_2Mo]_2$

2	Cp^-	12 electrons	– 2
4	CO	8 electrons	neutral
2	Mo(I)	<u>10 electrons</u>	+ 2
	Total	30 electrons	

$[2(18)-30]/2 = 3$ Mo–Mo bonds predicted.

3. m) (cont.)

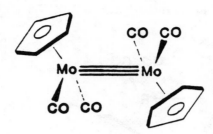

Klingler, R.J.; Butler, W.; Curtis, M.D. <u>J</u>. <u>Am</u>. <u>Chem</u>. <u>Soc</u>. **1975**, <u>97</u>, 3535.

n) $[\mu-H]_2-[Cp(NO)W(H)]_2$

2	Cp^-	12 electrons	– 2
2	H^-	4 electrons	– 2
2	NO^+	4 electrons	+ 2
2	$\mu-H^-$	4 electrons	– 2
2	W(II)	<u>8 electrons</u>	+ 4
Total		32 electrons	

and other isomers

$[2(18)-32]/2 = 2$ W–W bonds predicted

Legzdins, P.; Martin, J.T.; Oxley, J.C. <u>Organomet</u>. **1985**, <u>4</u>, 1263.

4. a) $Ni(CO)_4$, tetrahedral

4	CO	8 electrons	neutral
	Ni(0)	<u>10 electrons</u>	neutral
Total		18 electrons	

b) $(CO)_3Fe(COT)$

3	CO	6 electrons	neutral
	Fe(0)	8 electrons	neutral
	COT	<u>4 electrons</u>	neutral
Total		18 electrons	

4. c) CpCo(CO)$_2$

2	CO	4 electrons	neutral
	Cp$^-$	6 electrons	− 1
	Co(I)	8 electrons	+ 1
Total		18 electrons	

d) [Cp(CO)$_2$Fe]$_2$

4	CO	8 electrons	neutral
2	Cp$^-$	12 electrons	− 2
2	Fe(II)	12 electrons	+ 4
Total		34 electrons	plus 1 Fe-Fe bond

e) (CO)$_5$Re-H, octahedral

5	CO	10 electrons	neutral
	H$^-$	2 electrons	− 1
	Re(I)	6 electrons	+ 1
Total		18 electrons	

f) [Cp(CO)$_3$Mo]$_2$

6	CO	12 electrons	neutral
2	Cp$^-$	12 electrons	− 2
2	Mo(II)	8 electrons	+ 4
Total		34 electrons	plus 1 Mo-Mo bond

4. f) (cont.)

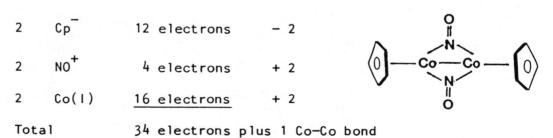

Other possibilities are $[Cp(CO)_2Mo]_2$ or $Cp_2Mo(CO)$

g) CpNi(NO)

Cp^-	6 electrons	− 1
NO^+	2 electrons	+ 1
Ni(0)	10 electrons	neutral
Total	18 electrons	

h) $[CpCo(NO)]_2$

2	Cp^-	12 electrons	− 2
2	NO^+	4 electrons	+ 2
2	Co(I)	16 electrons	+ 2
Total		34 electrons plus 1 Co—Co bond	

5. a) Tetrahedral **b)** Square planar **c)** Trigonal bipyramidal or octahedral if the complex is viewed as a metallacyclopropane **d)** Square planar **e)** Square planar **f)** Square planar **g)** Tetrahedral **h)** Square planar **i)** Square planar **j)** Distorted trigonal bipyramidal due to restrictions on the bond angles imposed by the chelating ligands.

6. Both structures **1** and **2** contain two central allyl protons which should have a chemical shift in the NMR spectrum between 4 and 6 ppm, not at 2.2 ppm, so these two structures can be eliminated on this basis. The assignments of proton resonances for **3** are:

5.2 (s, 10 H, Cp)

5.0 (m, 4 H, H_a)

3.3 (m, 4 H, H_b)

2.2 (m, 2 H, H_c)

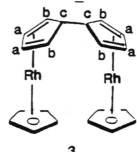

3

Electron counts are as follows:

For **1** and **2**, the electron count is the same:

	2	Cp^-	12 electrons	− 2
	2	$[\eta^3-allyl]^-$	8 electrons	− 2
	2	Rh(II)	14 electrons	+ 4
		Rh−Rh	2 electrons	neutral
Total			36 electrons	

For **3**:

	2	Cp^-	12 electrons	− 2
	2	diene	8 electrons	neutral
	2	Rh(I)	16 electrons	+ 2
Total			36 electrons	

Fischer, E.O.; Wawersik, H. _J. Organomet. Chem._ **1966**, _5_, 559.

7. a) $W(CO)(detc)_2(HC\equiv CH)$

CO	$2\ e^-$	neutral
2 detc	$8\ e^-$	− 2
W(II)	$4\ e^-$	+ 2
HC≡CH	$4\ e^-$	neutral
Total	18 electrons	

Acetylene donates 4 electrons.

b) $CpW(CO)(CH_3)(HC\equiv CH)$

CO	$2\ e^-$	neutral
Cp^-	$6\ e^-$	− 1
CH_3	$2\ e^-$	− 1
W(II)	$4\ e^-$	+ 2
HC≡CH	$4\ e^-$	neutral
Total	18 electrons	

Acetylene donates 4 electrons.

7. c) $Cp_2Mo(HC\equiv CH)$

  2 Cp⁻   12 e⁻ − 2

    Mo(II)   4 e⁻ + 2

    HC≡CH   2 e⁻ neutral

  Total   18 electrons

  Acetylene donates 2
  electrons.

d) $Mo(detc)_2(HC\equiv CPh)_2$

  2 detc   8 e⁻ − 2

    Mo(II)   4 e⁻ + 2

  2 HC≡CPh   6 e⁻ neutral

  Total   18 electrons

  Each phenylacetylene
  donates 3 electrons.

Templeton, J.L.; Ward, B.C. J. Am. Chem. Soc. **1980**, _102_, 3288.

8. The bonding in a carbene complex is depicted below.

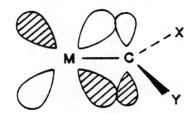

The pi bond between the metal and the carbon atoms arises as a result of a backbonding interaction between a filled metal d orbital of the appropriate symmetry and the empty carbene p orbital. Any factor that increases the electron density at the metal will increase the strength of the M—C pi bond, resulting in a higher activation energy for rotation around that bond. Aspects to consider include the d-electron count of the metal, the donor properties of the ligands, and in what row the metal is located.

a) In these two complexes, the only difference is the metal. The contrast between metals of different rows is apparent, with the rotational barrier increasing more than 2.3 kcal/mole on moving from second row molybdenum to third row tungsten. A possible explanation is that the electron density at a given metal atom increases down a column, thus the backbonding is increased and the metal-carbon pi bond is stronger.

b) In these two complexes, the only difference is the ligand. Because triethylphosphine is a better electron donor than triphenylphosphine, the complex with triethylphosphine as a ligand will have a stronger M—C pi bond.

c) Both Fe(II) and Re(I) are d^6 metals, the iron with strong electron-donating ligands and the rhenium with one electron-donating ligand and one electron-withdrawing ligand. If the electronic properties of the ligand were the only operative factor, we would predict that the iron complex would have a <u>higher</u> $\Delta G^{\ddagger}_{rot}$ than the rhenium complex. Since the observed trend is opposite, ligand effects must <u>not</u> be the major contributing factor. Thus, the increased electron density of a third row metal as compared to a first row metal explains the observed trends.

Brookhart, M.; Tucker, J.R.; Husk, G.R. <u>J</u>. <u>Organomet</u>. <u>Chem</u>. **1980,** <u>193</u>, C23.

Wong, W.K.; Tam, W.; Gladysz, J.A. <u>J</u>. <u>Am</u>. <u>Chem</u>. <u>Soc</u>. **1979,** <u>101</u>, 5440.

Kegley, S.E.; Brookhart, M.; Husk, G.R. <u>Organomet</u>. **1982,** <u>1</u>, 760.

9. The bonding in a metal carbonyl complex is depicted below.

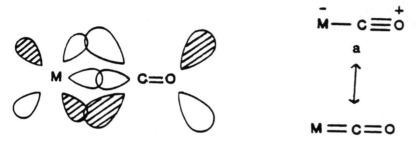

The pi bond between the metal and the carbon atoms arises as a result of a backbonding interaction between a filled metal d orbital of the appropriate symmetry and the empty pi star orbital of the carbonyl. Any factor that increases the electron density at the metal will increase the strength of the M—C pi bond, and thus the contribution from resonance form <u>b</u>. The result is a decrease in the CO bond order and a resulting decrease in the CO stretching frequency in the infrared spectrum of the complex. Aspects to consider include the d-electron count of the metal, the donor properties of the ligands, and the charge on the complex.

a) Both metals are d^6, but since $[W(CO)_5Cl]^-$ is anionic, it will have a lower CO stretching frequency.

b) The d^8 $Fe(CO)_5$ has more electron density at the metal center than the d^6 $Fe(CO)_4Br_2$ because of the lower oxidation state of the metal, thus $Fe(CO)_5$ will have a lower CO stretching frequency.

c) Both metals are d^6, but addition of phosphine ligands will increase the electron density at the metal. Thus, the CO stretch for $Mo(CO)_4(PPh_3)_2$ will be at a lower frequency than that for $Mo(CO)_6$.

d) The only difference between these two complexes is the nature of the ligands. Since trimethylphosphine is a better donor than

9. d) (cont.) triphenylphosphine, $Mo(CO)_4(PMe_3)_2$ will have a lower stretching frequency.

e) Both metals are d^8, but since $[Cp(CO)_2Fe]^-$ is anionic, it will have the lower CO stretching frequency.

10. For both **a)** and **b)**, there are only two orbitals of the appropriate symmetry to interact with the olefin pi star orbitals, the b_2 and the lowest energy b_1. The metal orbitals of <u>a</u> symmetry can only interact with the ethylene pi orbital and therefore have no preference for the orientation of the ethylene ligand.

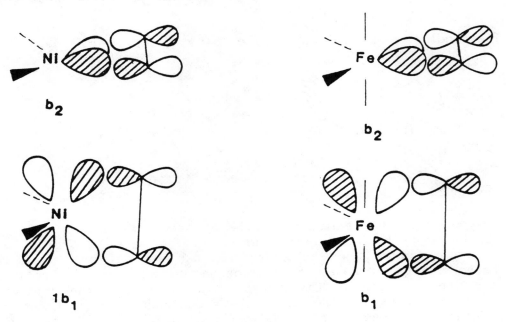

In order to distinguish which orientation will give the lower overall energy, it is useful to draw interaction diagrams (See Figures 2-1 and 2-2 on the following page for the NiL_2 case.

For both iron and nickel, the $1b_1$ orbital on the metal is lower in energy than the b_2 orbital, and thus interacts less with the relatively high-energy ethylene orbital of the same symmetry, since in general, the amount of bonding interaction between two orbitals is dependent on how close in energy they are. Thus, the lowest energy conformation will be that arising from interaction of the b_2 orbital on the metal with the ethylene ligand. In addition, the b_2 extends farther out in space and so has better overlap with the ethylene pi star orbitals, although this fact is not obvious from the information given.

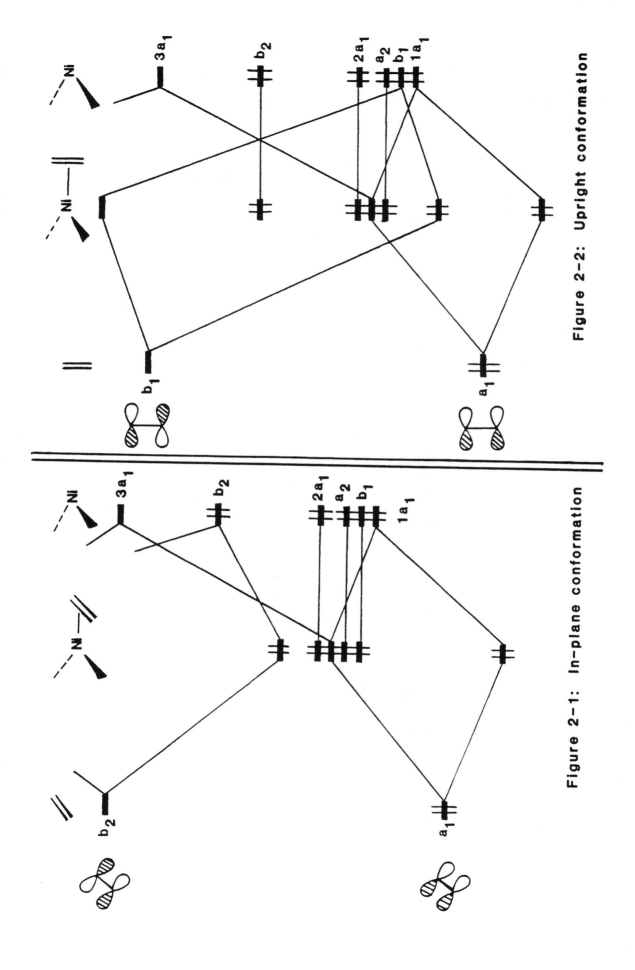

Figure 2-2: Upright conformation

Figure 2-1: In-plane conformation

10.   c) Electronically, either answer is correct. The barrier to rotation in Zeiss' salt is mainly steric and the observed orientation is perpendicular to the plane of the three chlorines, in order to minimize interactions between ethylene and the chlorines.

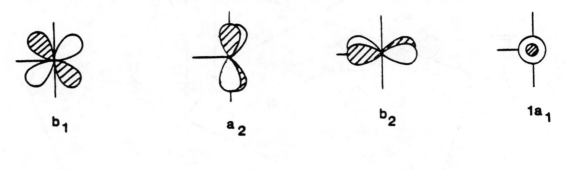

b_1 a_2 b_2 $1a_1$

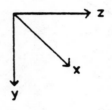

The a_1 orbital does not have a rotational preference, the b_1 orbital would prefer the in-plane geometry, and the b_2 orbital would prefer the observed out-of-plane geometry. The b_1 and b_2 orbitals have similar energies and hybridizations, so both can bond equally well to the ethylene. With the in-plane geometry, there is a significant steric interaction between the chlorines and the ethylene; therefore the out-of-plane geometry is preferred.

b_2

Albright, T.A.; Hoffmann, R.; Thibeault, J.C.; Thorn, D.L. J. Am. Chem. Soc. **1979**, 101 3801.

11. a) $CoBr_2(CO)L_2$ is a 17-electron species.

2	Br^-	4 electrons	-2
	CO	2 electrons	neutral
2	L	4 electrons	neutral
	Co(II)	<u>7 electrons</u>	$+2$
Total		17 electrons	

b) The CO stretching frequencies are indicative of the strength of the carbon-oxygen bond, and of the electron donating ability of the donor ligand. There are two resonance forms for the bonding of a metal-carbonyl:

$$M \!\!=\!\!= C \!\!=\!\!= O \qquad \longleftrightarrow \qquad M \!\!-\!\! C \!\!\equiv\!\! O$$

$$I \qquad\qquad\qquad\qquad\qquad II$$

A low CO stretching frequency is indicative of a significant amount of electron density between the metal and the carbonyl carbon and a significant amount of backbonding into the pi star orbital of the carbon monoxide (resonance structure I). This situation arises when the metal has electron-rich ligands. A metal with electron-deficient ligands would be more accurately represented by resonance structure II, for which a higher CO stretching frequency would be expected. For this reaction, electronic effects appear to be important if we compare complexes with ligands of identical steric requirements (cone angles). For example, for $P(n-Pr)_3$ and PEt_2Ph, K_d increases with an increase in the CO stretching frequency.

Nevertheless, the dominant factor in these reactions appears to be the steric requirements of the phosphine ligands. The observed trend for the dissociation constants for these cobalt complexes correlates very well with cone angle, with a marked increase in K_d with an increase in the cone angle. Two causes of this effect are:

1) Release of strain energy upon loss of a ligand.

2) A secondary electronic effect which results from a longer M-P bond (and thus, less electron donation from the phosphine to the metal) because of the steric requirements of the groups on the phosphine ligand.

Tolman, C.A. Chem. Rev. **1977**, 77, 313.

Tolman, C.A.; Seidel, W.C.; Gosser, L.W. J. Am. Chem. Soc. **1974**, 96, 53.

Bressan, M.; Corain, B.; Rigo, P.; Turco, A. Inorg. Chem. **1970**, 9, 1733.

12.   The proper approach for this problem is to:

1) Choose the correct molecular orbital diagram based on the geometry of the metal fragment.

2) Fill the metal orbitals with the appropriate number of d electrons.

3) Separate the relevant orbitals from the others, disregard the other orbitals, and compare the resulting MO diagram to the hydrocarbon fragment MO diagrams. (In the solutions that follow, the relevant orbitals are separated from the other orbitals by dotted lines.)

4) Choose the appropriate hydrocarbon fragment, supply enough electrons to match the d electron configuration of the metal, and assign a charge to the carbon fragment.

5) If all has gone well, get a beer.

a) The $[PtCl_3]^-$ fragment is diagram D, d^8. CH_3^+ is isolobal. With the ML_3 (D) fragment, there is an additional nonbonding orbital, $1a_1$ (similar to a dz^2), that arises when this fragment is formed by removal of two axial (or trans) ligands from an ML_5 system. Removal of these ligands decreases the energy of the a_1 orbital and it becomes part of the nonbonding set. Thus for system D, there are <u>four</u> nonbonding orbitals that become unimportant in the isolobal analysis.

12. b) The $Fe(CO)_3$ fragment is diagram C, d^8. CH^+ is isolobal.

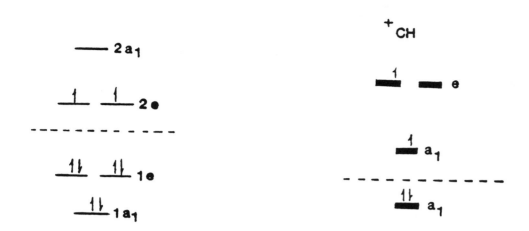

Note that the isolobal analogy does not imply that the <u>ordering</u> of the energy levels is the same for the metal fragment as for the carbon fragment; instead it tells us only that the number of available orbitals and electrons is the same. For both $Fe(CO)_3$ and ^+CH, there are two electrons and three available orbitals.

c) The $(CO)_4Fe$ fragment is diagram B, d^8. $:CH_2$ is isolobal.

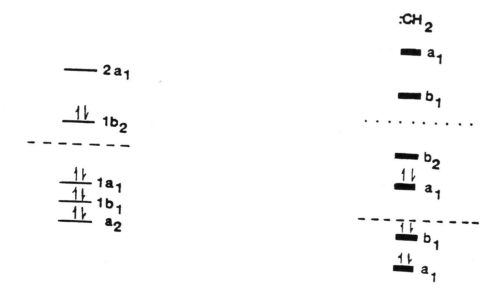

12.   d) The Cp(CO)Rh fragment is diagram B, d^8.   :CH$_2$ is isolobal.

e) The $[PdL_2]^{+2}$ fragment is diagram E, d^8. CH_2^{+2} is isolobal. With the ML_2 (E) fragment, there is an additional nonbonding orbital, $2a_1$ (similar to a dz^2), that arises when this fragment is formed by removal of two axial (or trans) ligands from an ML_4 system (Diagram B). Removal of these ligands decreases the energy of the $2a_1$ orbital and it becomes part of the nonbonding set.   Thus for system E, there are <u>four</u> nonbonding orbitals that become unimportant in the isolobal analysis.

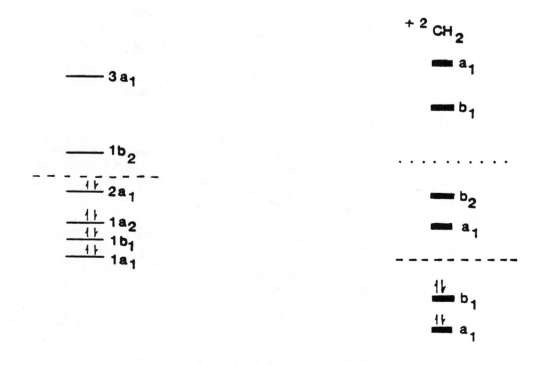

12. f) The [FeCp]$^+$ fragment is diagram C, d^6. CH^{+3} is isolobal.

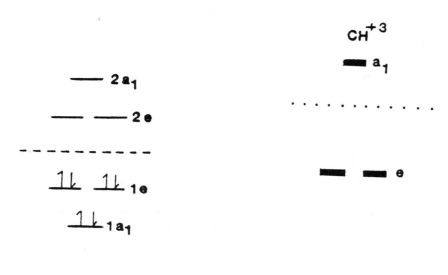

g) The Os(CO)$_4$ fragment is diagram B, d^8. :CH$_2$ is isolobal.

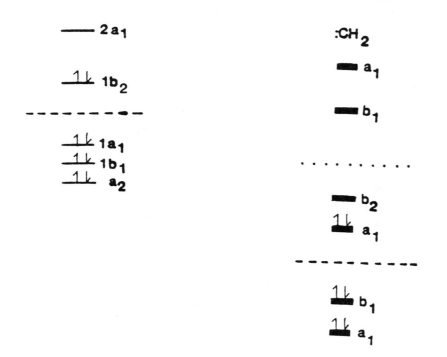

12. h) The Re(CO)$_5$ fragment is diagram A, d^7. $\cdot$CH$_3$ is isolobal.

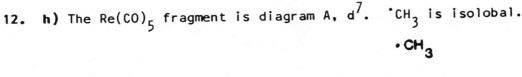

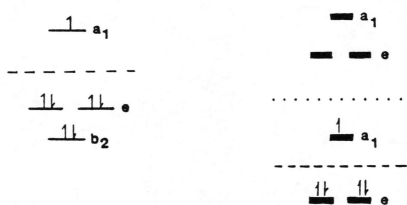

i) The Ni(CO)$_2$ fragment is diagram E, d^{10}. :CH$_2$ is isolobal. With the ML$_2$ (E) fragment, there is an additional nonbonding orbital, 2a$_1$ (similar to a dz^2), that arises when this fragment is formed by removal of two axial (or trans) ligands from an ML$_4$ system (Diagram B). Removal of these ligands decreases the energy of the 2a$_1$ orbital and it becomes part of the nonbonding set. Thus for system E, there are <u>four</u> nonbonding orbitals that become unimportant in the isolobal analysis.

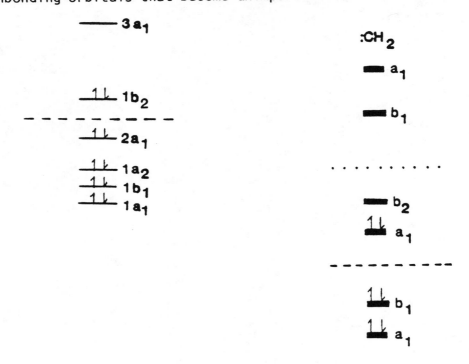

12. (cont.) Note that most of these examples contain metals with a high d-electron count. These systems were chosen because the splitting between the bonding and nonbonding orbitals is large and thus electrons are necessarily added to the lower levels first. When the gap between nonbonding and bonding orbitals is relatively small, as for metals with a lower d-electron count, a number of possibilities exist for filling up the orbitals. For more information, see Hoffmann, R. Angew. Chem. Int. Ed. Engl., **1982**, 21, 711.

Other useful references are:

Elian, M.; Hoffmann, R. Inorg. Chem. **1975**, 14, 1058.

Elian, M.; Chen, M.M.L.; Mingos, D.M.P.; Hoffmann, R. Inorg. Chem. **1976**, 15, 1148.

Hoffmann, R. Science **1981**, 211, 995.

Pinhas, A.R.; Albright, T.; Hofmann, P.; Hoffmann, R. Helv. Chim. Acta. **1980**, 63, 29.

13. The complex that was actually studied experimentally was

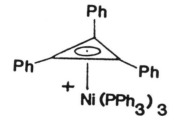

and the one studied theoretically was

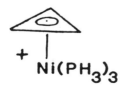

13. (cont.) The molecular orbital interaction diagram is shown below.

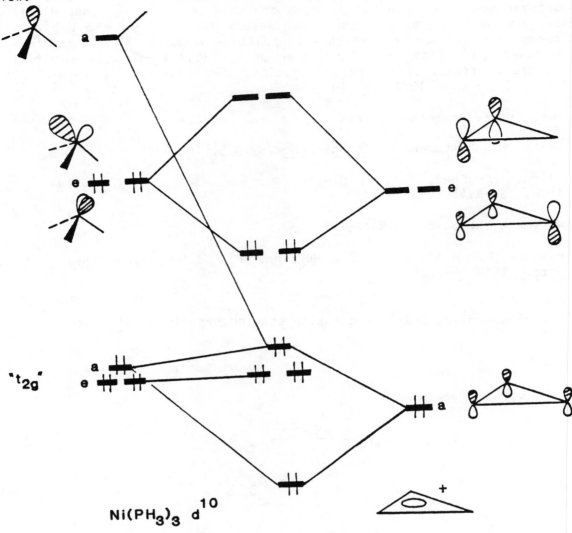

$$Ni(PH_3)_3 \; d^{10}$$

a) The major interaction is between the filled ML_3 _e_ set and the empty cyclopropenyl _e_ set. These orbitals also determine the relative orientation.

b) Looking at the symmetric set of orbitals, it is clear that maximum overlap is obtained by conformation _2_.

vs

_1_   _2_
                                 preferred

13. (cont.) The antisymmetric set of orbitals also prefers the same relative orientation of the two C_3 rotors.

vs

<u>1</u>

<u>2</u>

preferred

Thus the relative orientation (from a top view of the molecule) is shown below.

Mealli, C.; Midollini, S.; Moneti, S.; Sacconi, L.; Silvestre, J.; Albright, T.A. J. Am. Chem. Soc. **1982**, 104, 95.

Jemmis, E.D.; Hoffmann, R. J. Am. Chem. Soc. **1980**, 102, 2570.

14. In the formyl complex, the best pi bonding between the p orbital of the formyl ligand and a metal d orbital will occur when the p orbital is aligned with the higher energy orbital, I, because of the better donor properties of triphenylphosphine compared to nitrosyl. Thus, the formyl ligand in this rhenium complex is oriented such that it is aligned with the nitrosyl group as shown below.

Tam, W.; Lin, G.-Y.; Wong, W.-K.; Kiel, W.A.; Wong, V.K.; Gladysz, J.A. J. Am. Chem. Soc. **1982**, 104, 141.

Schilling, B.E.R.; Hoffmann, R.; Faller, J.W. J. Am. Chem. Soc. **1979**, 101, 592.

15. The molecular orbital interaction diagram for $Fe(CO)_3$ and trimethylenemethane is shown below.

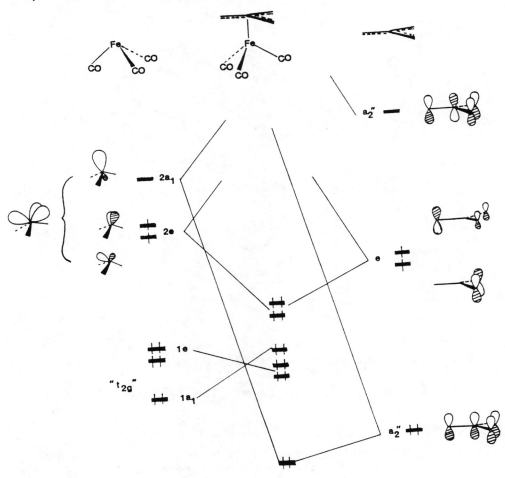

The preferred orientation is <u>2</u>, since maximum orbital overlap is obtained in this conformation.

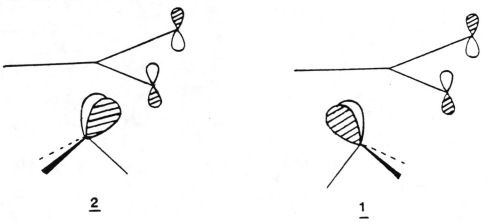

Albright, T.A.; Hofmann, P.; Hoffmann, R. <u>J</u>. <u>Am</u>. <u>Chem</u>. <u>Soc</u>. **1977**, <u>99</u>, 7546.

3

Ligand Substitution
Reactions

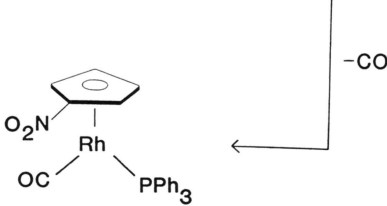

3: LIGAND SUBSTITUTION REACTIONS

QUESTIONS

1. Predict reasonable products for the following reactions.

a)

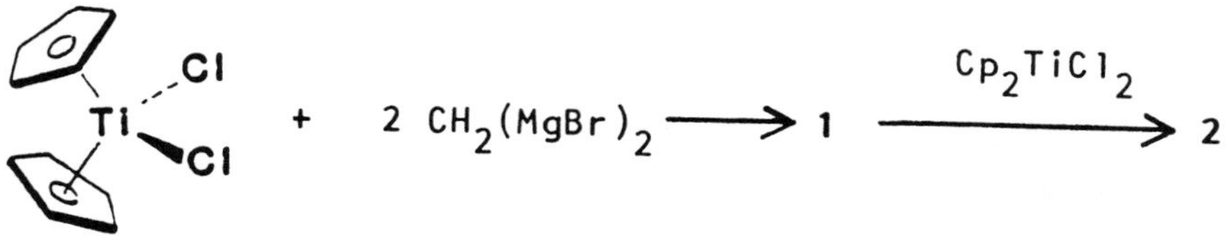

$$Cp_2TiCl_2 + 2\ CH_2(MgBr)_2 \longrightarrow 1 \xrightarrow{Cp_2TiCl_2} 2$$

b)

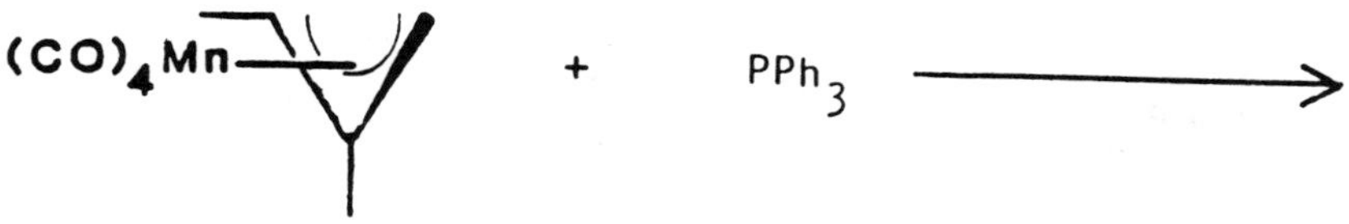

$$(CO)_4Mn{-} \qquad + \qquad PPh_3 \longrightarrow$$

c)

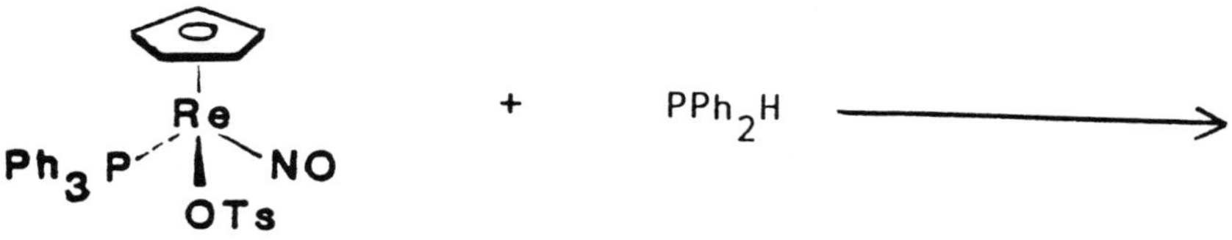

$$+ \qquad PPh_2H \longrightarrow$$

d)

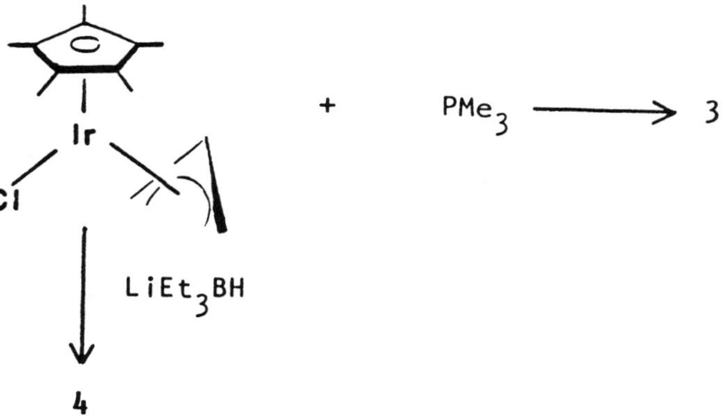

$$+ \qquad PMe_3 \longrightarrow 3$$

LiEt$_3$BH

4

e)

$$+ \qquad PPh_3 \xrightarrow{Me_3NO}$$

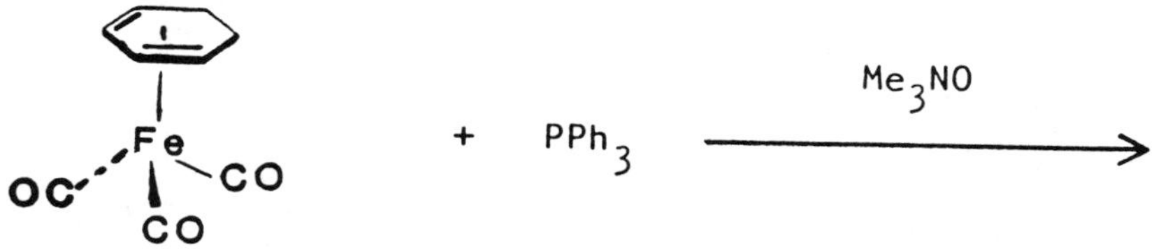

1. f)

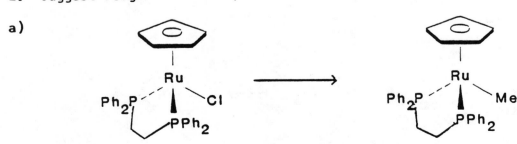

g)

h) $PdCl_2$ $\xrightarrow{\text{reduction}}$ $Pd(0)$

i)

$Cr(CO)_6$ +

j) $Co(CO)_3(NO)$ + PPh_3 $\xrightarrow{h\nu}$

k) $MoCl_3(THF)_3$ +   excess PMe_3 $\longrightarrow$

l) $Mo(CO)_6$ + $Me_2PCH_2CH_2P(Ph)CH_2CH_2PMe_2$ $\xrightarrow{\Delta}$

m) $Mo(CO)_6$ + CH_3CN $\xrightarrow{\Delta}$

n) $NiCl_2$ + $Me_2PCH_2CH_2PPh_2$ $\longrightarrow$

2.   Suggest reagents to carry out the following transformations.

a)

2. b)

c)

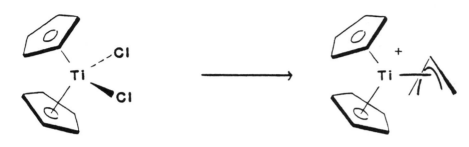

d)

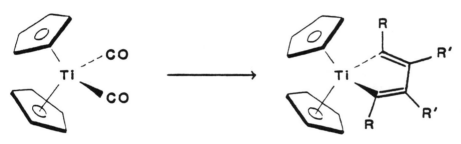

e) $MoCl_4(THF)_2 \longrightarrow MoCl_3(THF)_3$

f)

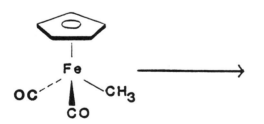

g)

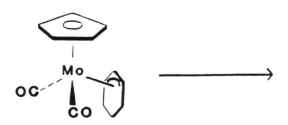

3.   Provide   metal-containing   products   for   each   of   the   following
reactions:

a) 1,3-cyclohexadiene   + $Fe(CO)_5$ $\xrightarrow{h\nu}$

b) $Co_2(CO)_8$ +   diphenylacetylene $\longrightarrow$

c) $NaMn(CO)_5$ + $CH_2=CH-CH_2Cl$ $\longrightarrow$

d) $Cr(CO)_6$ $\xrightarrow[\Delta]{\text{benzene}}$

e) 3,4-dichlorocyclobutene   + $Ni(CO)_4$ $\longrightarrow$
HINT
This is a dimeric product.

4.   A mixture of $Fe(CO)_5$ and $Mn_2(CO)_{10}$ dissolved in hexane was
irradiated for 45 minutes with ultraviolet light from a mercury lamp.
Volatile liquids were then removed under vacuum and the solid residue
sublimed onto a cold finger as red crystals.   This compound (5) was
quite soluble in organic solvents; its IR spectrum showed bands at
2067, 2019, and 1987 cm^{-1}.   The mass spectrum of 5 showed a series of
ions separated by 28 mass units starting at m/e 166 and proceeding to
m/e 558. Significant amounts of other series of ions beginning at m/e
55, 56, and 111 were observed, but not from m/e 110. [Note: ^{55}Mn and
^{56}Fe are the dominant isotopes for manganese and iron, respectively.]

a) Formulate a structure for 5 and relate it to the data given.
b) Show that your structure conforms to the 18 electron rule.

5.   The following reactions occur:

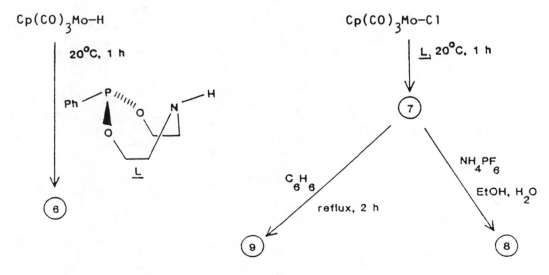

5. (cont.) Postulate structures for **6, 7, 8,** and **9** based on the data.

Complex	^{1}H NMR	selected IR data(cm^{-1})
6	3.44 (m, 4 H) 3.96 (m, 4 H) 5.26 (s, 5 H) 7.49 (m, 5 H) −6.41 (d, J = 65 Hz, 1 H)	3389 1945 1860
7	3.44 (m, 4 H) 4.01 (m, 4 H) 5.29 (s, 5 H) 7.50 (m, 5 H)	3355 1966 1880
8	3.21 (m, 4 H) 3.96 (m, 2 H) 4.64 (m, 2 H) 6.00 (s, 5 H) 7.42 (m, 1 H) 7.70 (m, 5 H)	3380−3400 1990 1900
9	3.38 (m, 4 H) 4.41 (m, 4 H) 5.06 (d, J = 3 Hz, 5 H) 5.72 (m, 1 H) 7.48 (m, 5 H)	3160 1800

HINTS
1) Complex **8** is cationic.
2) Reaction of **6** with chloroform gives **7**.
3) N−H protons often appear as broad singlets which may be obscured by other proton resonances.
4) All complexes are 18−electron species.

6. Propose a mechanism for the following substitution reaction.

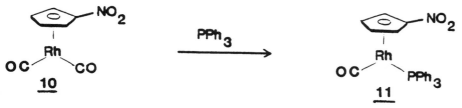

Your mechanism must be consistent with the following three observations:

1) Using excess PPh_3, the rate is first order in rhodium complex.
2) Plots of k_{obs} vs $[PPh_3]$ are linear with a zero intercept.
3) $\Delta S^{\ddagger}$ = −17.9 eu

7.   Two different mechanisms have been proposed to explain the following reaction.

$$Re_2(CO)_{10} + L \longrightarrow Re_2(CO)_9L + CO$$

Mechanism I

$$Re_2(CO)_{10} \rightleftharpoons Re_2(CO)_9 + CO$$

$$\downarrow +L$$

$$Re_2(CO)_9L$$

Mechanism II

$$Re_2(CO)_{10} \rightleftharpoons 2 \cdot Re(CO)_5$$

$$\cdot Re(CO)_5 + L \longrightarrow \cdot Re(CO)_4L + CO$$

$$\cdot Re(CO)_5 + \cdot Re(CO)_4L \longrightarrow Re_2(CO)_9L$$

Using the fact that there are two isotopes of rhenium, ^{185}Re and ^{187}Re, how could you distinguish between these two mechanisms? What experimental technique(s) would be required?

8. Predict the products of the following reactions.

1)

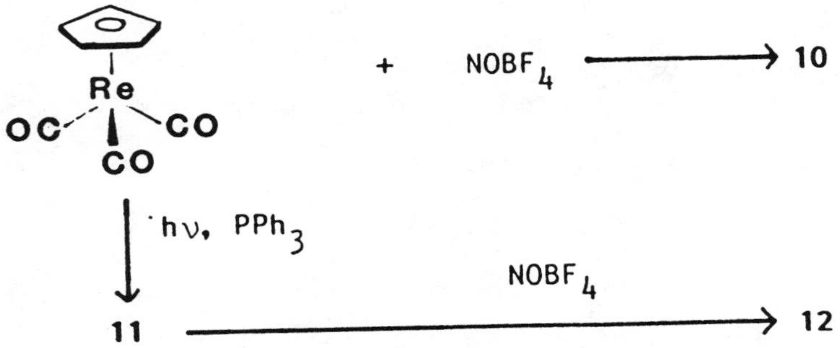

2)

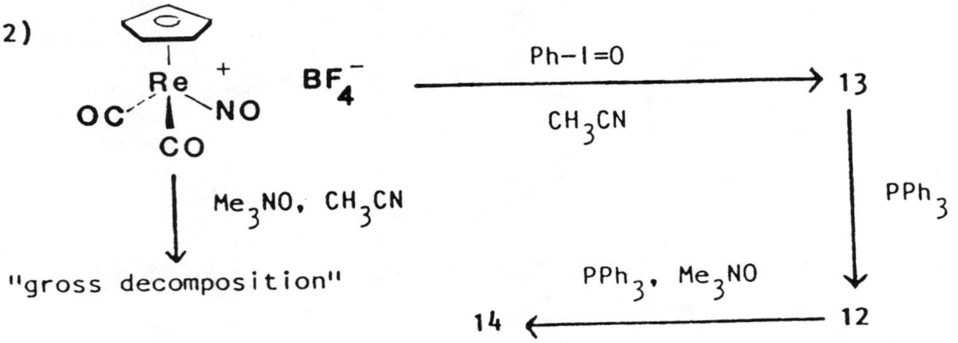

8. (cont.)

3)

$$\text{[Cp Re(CO)(NO)(PPh}_3\text{)]}^+ \text{ BF}_4^- \quad + \quad \text{LiEt}_3\text{BH} \longrightarrow \text{15}$$

4)

$$\text{15} \quad + \quad \text{[Cp Re(NO)(PPh}_3\text{)(=C(OMe)H)]}^+ \quad \xrightarrow{-70^\circ C} \quad \text{12} \quad + \quad \text{16}$$

^{1}H NMR data for 15

7.50–7.36 (m, 15 H)
5.25 (s, 5 H)
16.48 (s, 1 H)

IR data for 15

1560 cm^{-1}

^{1}H NMR data for 16

7.40–7.25 (m, 15 H)
5.04 (d, J = 0.5 Hz, 5 H)
5.09 (d of d, J = 10.5, 5.6 Hz, 1 H)
5.45 (d of d, J = 10.5, 2.0 Hz, 1 H)
3.16 (s, 3 H)

9. Let $A = [(\eta^6\text{-}C_6H_6)Mn(CO)_3]^+$ and $L = PBu_3$.

$$A \quad + \quad L \quad \underset{k_{-1}}{\overset{k_1}{\rightleftharpoons}} \quad B \qquad\qquad (1)$$

$$B \quad \xrightarrow{h\nu} \quad C \qquad\qquad (2)$$

Spectral Data	IR(cm^{-1})	^{1}H NMR(δ, acetone-d$_6$)
A	2080, 2026	6.90
B	2028, 1950	6.30 (1 H), 5.50 (2 H), 4.40 (1 H), 3.40 (2 H), plus PBu$_3$ resonances
C	1997, 1950	6.42 plus PBu$_3$ resonances

a) Identify B and **C**.
b) Kinetic data were obtained for the approach to equilibrium for reaction (1) with L in excess. The rate expression is shown in equation (3).

$$-d[A]/dt = (k_1[L] + k_{-1})([A] - [A]_\infty) \qquad (3)$$

or in a simplified form,

9. (cont.)

$$-d[A]/dt = k_{obs}([A] - [A]_\infty) \qquad (4)$$

where $k_{obs} = k_1[L] + k_{-1}$

Plots of $\ln([A]-[A]_\infty)$ versus time were linear. Derive the rate expression in equation (3) and show the integrated form of this rate expression that was plotted to obtain the straight lines.

c) Calculate ΔH^o, ΔS^o, and K_{eq} at $25^o C$ based on the data given below.

T (^{o}C)	k_1 ($s^{-1}M^{-1}$)	k_{-1} (s^{-1})
10	480	0.55
20	770	1.8
33	1420	7.7

10. The following reaction is observed, with ligand substitution occurring before dimerization takes place.

$$2 \cdot (CO)_3 MnL_2 \; + \; CO \; \longrightarrow \; Mn_2(CO)_8 L_2 \; + \; 2\,L$$

The following observations have been made:
1) The pseudo-first-order rate constant is approximately proportional to $[CO]$.
2) Added L has no effect on the rate of the reaction.
3) The rate depends on the nature of L, with $k_{obs} = 0.32\ M^{-1}s^{-1}$ for L = $P(i-Bu)_3$ and $k_{obs} = 42\ M^{-1}s^{-1}$ for L = $P(n-Bu)_3$.

Propose a mechanism for the reaction, classify it as associative or dissociative, and derive the rate expression that is consistent with all observations. Clearly indicate which step is rate-determining.

11. In ten words or less, state why life exists.

ANSWERS

1. The answers here represent the actual products for the reactions as published in the literature; however, there may be other products that could reasonably be expected from the reactions given.

a)

1 = 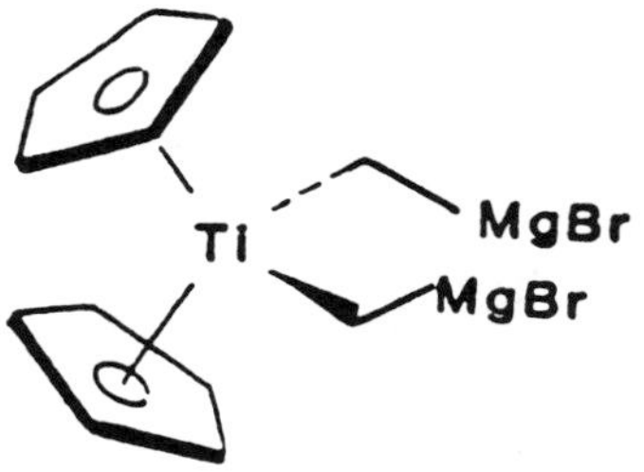   2 =

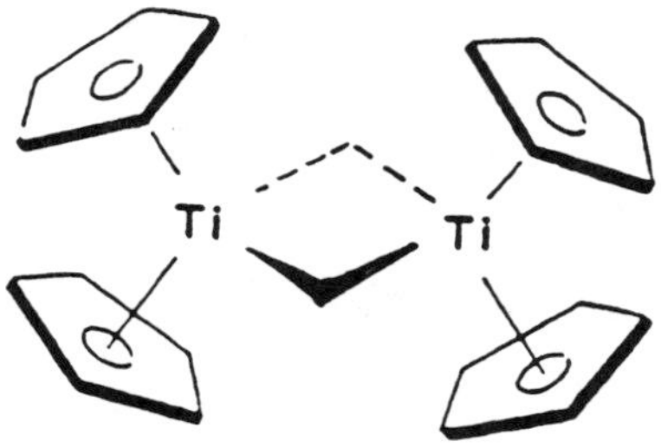

Van de Heisteeg, B.J.J.; Schat, G.; Akkerman, O.S.; Bickelhaupt, F. Organomet. **1985**, _4_, 1141.

b)

CO   + $(PPh_3)(CO)_3 Mn$—

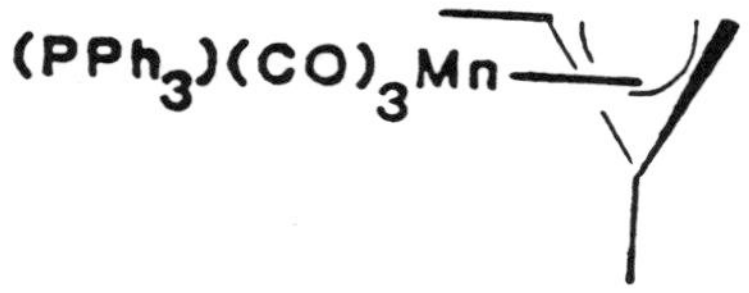

Palmer, G.T.; Basolo, F. _J._ **Am.** **Chem.** **Soc.** **1985**, _107_, 3122.

c)

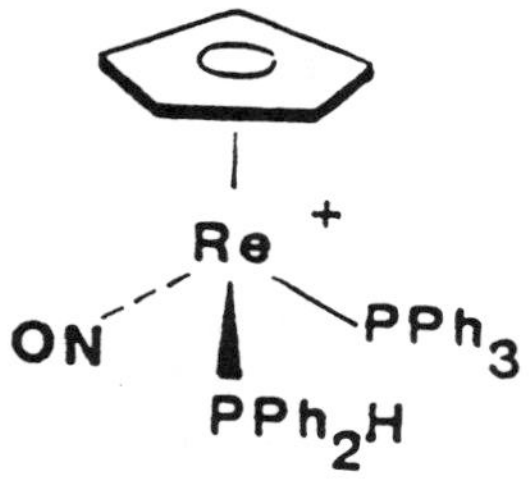

Buhro, W.E.; Georgiou, S.; Hutchinson, J.P.; Gladysz, J.A. _J._ **Am.** **Chem.** **Soc.** **1985**, _107_, 3346.

d)

3 =

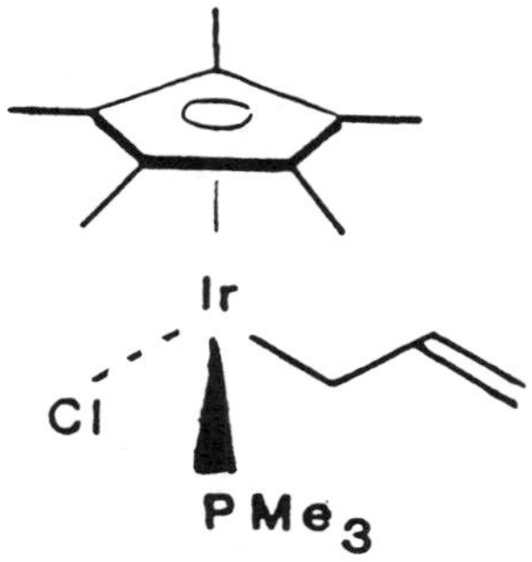

1. d) (cont.)

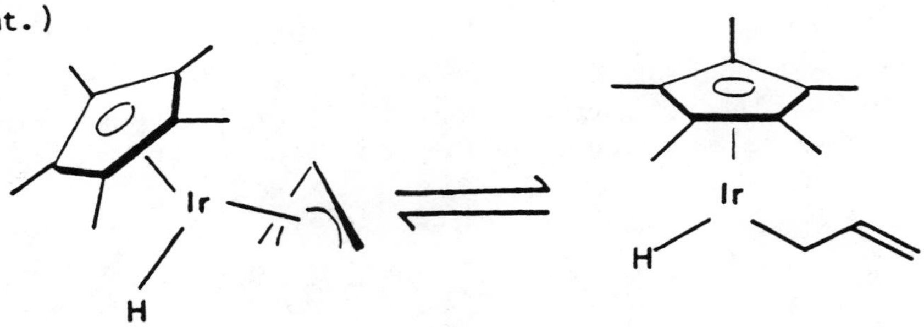

McGhee, W.D.; Bergman, R.G. J. Am. Chem. Soc. 1985, 107, 3388.

e)

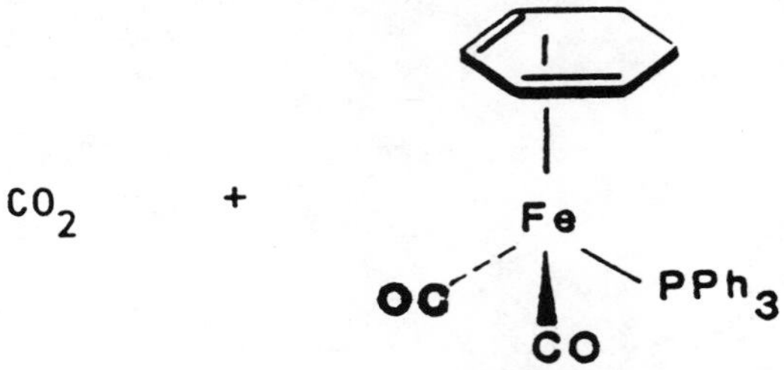

CO_2 +

Birch, A.J.; Kelly, L.F. J. Organomet. Chem. 1985, 286, C5.

f)

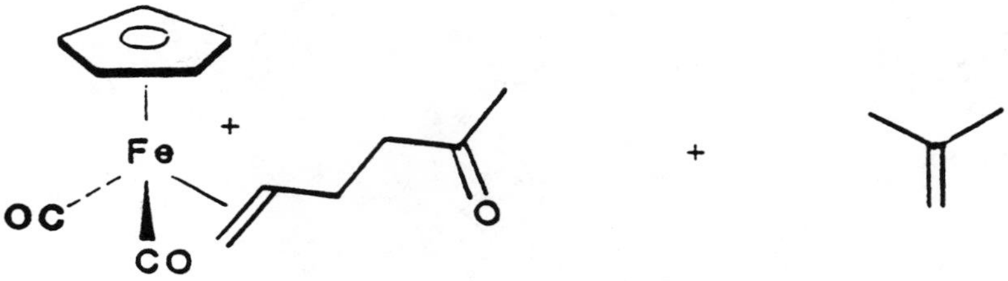

Berryhill, S.R.; Rosenblum, M. J. Org. Chem. 1980, 45, 1984.

g)

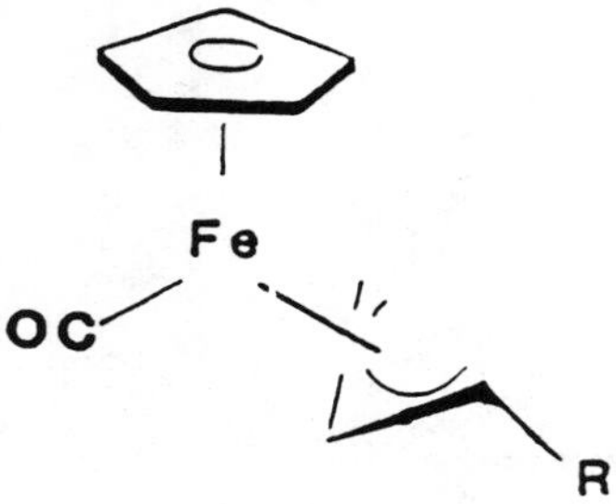

Green, M.L.H.; Nagy, P.L.I. J. Chem. Soc. 1963, 189.

h)

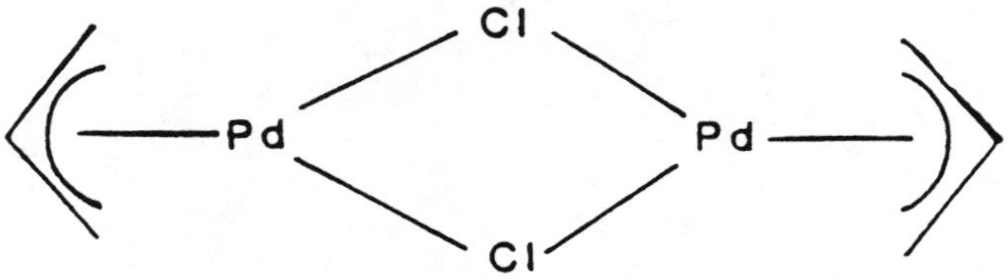

Trost, B.M. Tetrahedron 1977, 33, 2615.

1. i)

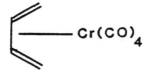

Geoffroy, G.L.; Wrighton, M.S. "Organometallic Photochemistry" (Academic Press, 1979), p. 176.

j) $Co(CO)_2(NO)(PPh_3)$

Sabherwal, I.H.; Burg, A. Chem. Comm. 1969, 853.

k) $MoCl_3(PMe_3)_3$

Carmona, E.; Marin, J.M.; Poveda, M.L.; Atwood, J.L.; Rogers, R.D. J. Am. Chem. Soc. 1983, 105, 3014.

l)

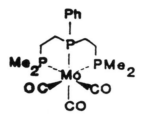

King, R.B.; Zinich, J.A.; Cloyd, J.C., Jr. Inorg. Chem. 1975, 14, 1554.

m) $Mo(CO)_3(CH_3CN)_3$

Tate, D.P.; Knipple, W.R.; Augl, J.M. Inorg. Chem. 1962, 1, 433.

n)

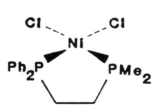

King, R.B.; Zinich, J.A.; Cloyd, J.C., Jr. Inorg. Chem. 1975, 14, 1554.

2. The answers here represent actual reagents used to carry out the transformations in a given literature reference. There may be other reagents that would effect the same transformations.

2. a) MeLi

Davies, S.G. "Organotransition Metal Chemistry: Applications to Organic Synthesis" (Pergamon Press, 1982), p. 25.

b) $AlCl_3$, CO

Fischer, E.O.; Fichtel, K. <u>Chem</u>. <u>Ber</u>. **1961**, <u>94</u>, 1200.

c) Allyl—MgBr

Cotton, F.A.; Wilkinson, G. "Advanced Inorganic Chemistry", 4th ed. (John Wiley and Sons, 1980), p. 705.

d) $2 R-C \equiv C-R'$

Cotton, F.A.; Wilkinson, G. "Advanced Inorganic Chemistry", 4th ed. (John Wiley and Sons, 1980), p. 706.

e) Zn, THF

Cotton, F.A.; Wilkinson, G. "Advanced Inorganic Chemistry", 4th ed. (John Wiley and Sons, 1980), p. 864.

f) PPh_3, Δ

Wojcicki, A. <u>Adv</u>. <u>in</u> <u>Organomet</u>. <u>Chem</u>. **1973**, <u>11</u>, 87.

g) $NOPF_6$

Faller, J.W.; Murray, H.H.; White, D.L.; Chao, K.H. <u>Organomet</u>. **1983**, <u>2</u>, 400.

3. a)

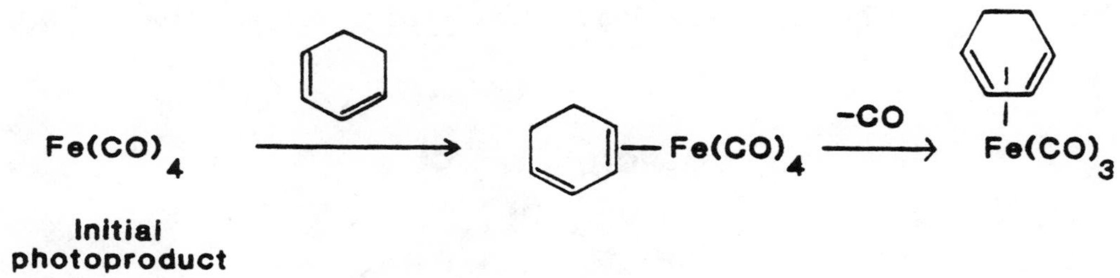

Hallam, B.F.; Pauson, P.L. <u>J</u>. <u>Chem</u>. <u>Soc</u>. **1958**, 642.

b)

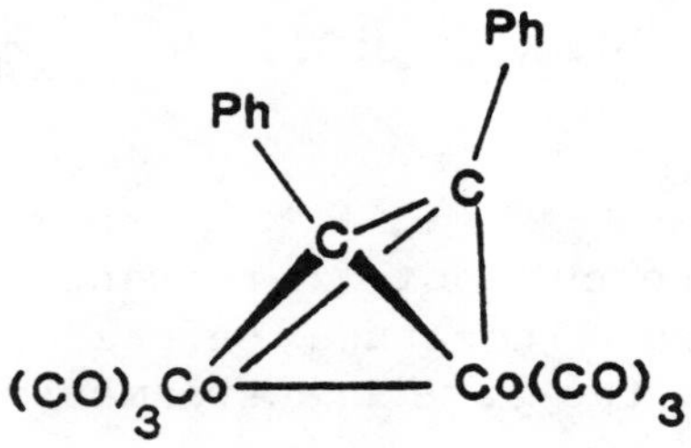

Sly, W.G. <u>J</u>. <u>Am</u>. <u>Chem</u>. <u>Soc</u>. **1959**, <u>81</u>, 18.

3. c)

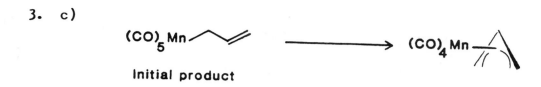

Initial product

Collman, J.P.; Hegedus, L.S. "Principles and Applications of Organotransition Metal Chemistry" (University Science Books, 1980), p. 137.

d)

Lukehart, C.M. "Fundamental Transition Metal Organometallic Chemistry" (Brooks/Cole Publishing Co., 1985), p. 128.

e)

Efraty, A. Chem. Revs. 1977, 77, 691.

4. a) The infrared spectral data indicate that only terminal carbonyls are present, thus all structures involving bridging carbonyl groups can be eliminated. The mass spectral data shows a molecular ion at 558 m/e, which loses carbonyl groups (28 m/e) one at a time, down to a mass of 166, thus a total of 392 m/e or 14 carbonyl groups are lost. The fragment at 111 m/e corresponds to a Mn–Fe unit. Thus, the composition of the 166 peak is two Mn atoms and one Fe atom. From these data, an empirical formula of $Mn_2Fe(CO)_{14}$ can be obtained. The fact that no 110 m/e fragment was observed suggests that there is no Mn–Mn bond in the molecule.

The only reasonable structure is shown on the following page.

4. (cont.)

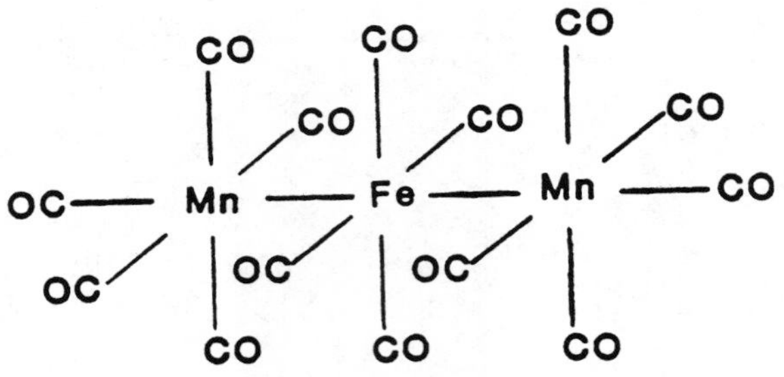

5

b) An X-ray structure of the compound proves this structure to be the correct one for **5**.

Abel, E.W.; McLean, R.A.N.; Moorhouse, S. Inorg. Nucl. Chem. Let. **1971**, 7, 587.

Evans, G.O.; Sheline, R.K.; Inorg. Chem., **1971**, 10, 1598.

Schubert, E.H.; Sheline, R.K. Z. Naturforsch. **1965**, b20, 1306.

5. The structures of complexes **6-9** are shown below.

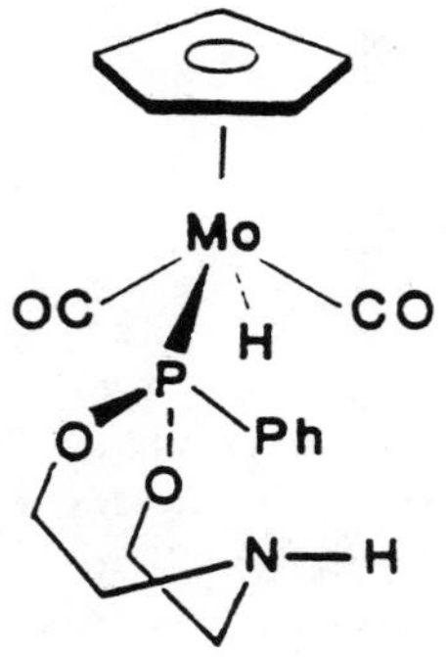

6

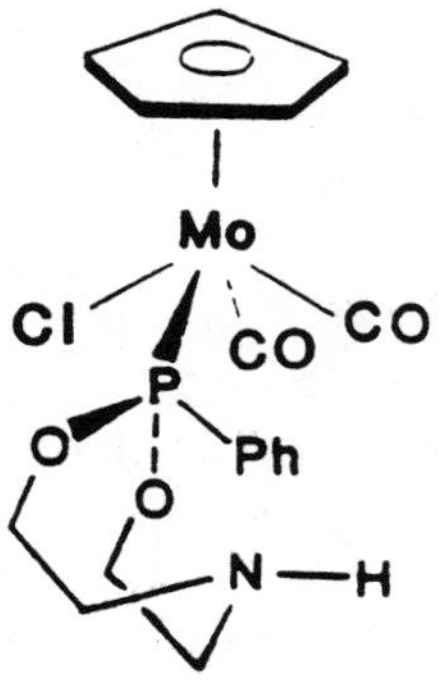

7

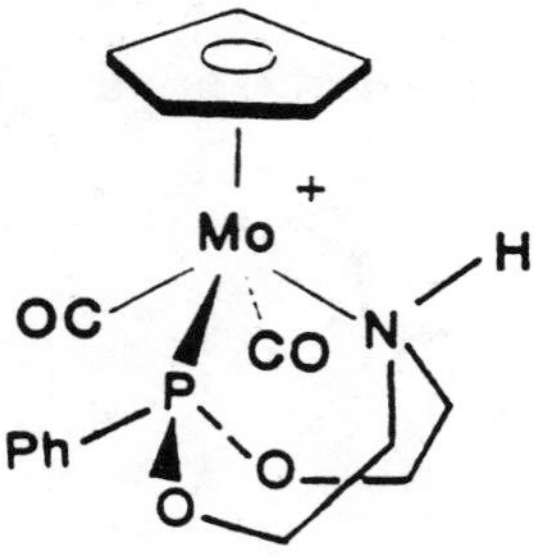

8

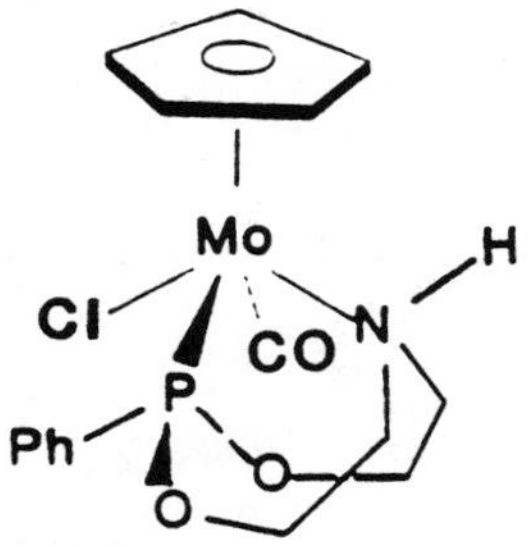

9

5. (cont.) The NMR and IR data are assigned as follows:

Complex	^{1}H NMR	selected IR data (cm^{-1})
6	3.44 (m, 4 H, $-CH_2N-$)	3389 (N–H stretch)
	3.96 (m, 4 H, $-CH_2O-$)	1945 (CO stretch)
	5.26 (s, 5 H, Cp)	1860 (CO stretch)
	7.49 (m, 5 H, Ph)	
	-6.41 (d, J_{P-H} = 65 Hz, 1 H, Mo–H)	
7	3.44 (m, 4 H, $-CH_2N-$)	3355 (N–H stretch)
	4.01 (m, 4 H, $-CH_2O-$)	1966 (CO stretch)
	5.29 (s, 5 H, Cp)	1880 (CO stretch)
	7.50 (m, 5 H, Ph)	
8	3.21 (m, 4 H, $-CH_2N-$)	3380–3400 (N–H stretch)
	3.96 (m, 2 H, $-CH_2O-$)	1990 (CO stretch)
	4.64 (m, 2 H, $-CH_2O-$)	1900 (CO stretch)
	6.00 (s, 5 H, Cp)	
	7.42 (m, 1 H, –NH)	
	7.70 (m, 5 H, Ph)	
9	3.38 (m, 4 H, $-CH_2N-$)	3160 (N–H stretch)
	4.41 (m, 4 H, $-CH_2O-$)	1800 (CO stretch)
	5.06 (d, J_{P-H} = 3 Hz, 5 H, Cp)	
	5.72 (m, 1 H, –N–H)	
	7.48 (m, 5 H, Ph)	

Points to note:
1) Upon coordination of the nitrogen atom to the metal, the N–H stretching frequency decreases dramatically and the number of metal carbonyl stretches decreases to one, as it must. Additionally, the N–H proton becomes visible in the NMR spectrum.

2) Upon formation of a cationic species, the Cp resonance shifts significantly downfield. In addition, the ethylene bridge hydrogens become nonequivalent, presumably because the structure is more rigid. The N–H proton is also visible in the NMR spectrum of this species. The CO stretches are shifted to higher energy, again because of the positive charge on the metal.

Wachter, J.; Jeanneaux, F.; Riess, J.G. Inorg. Chem. **1980**, 19, 2169.

6. There are two possible mechanisms.

1) Dissociative Mechanism

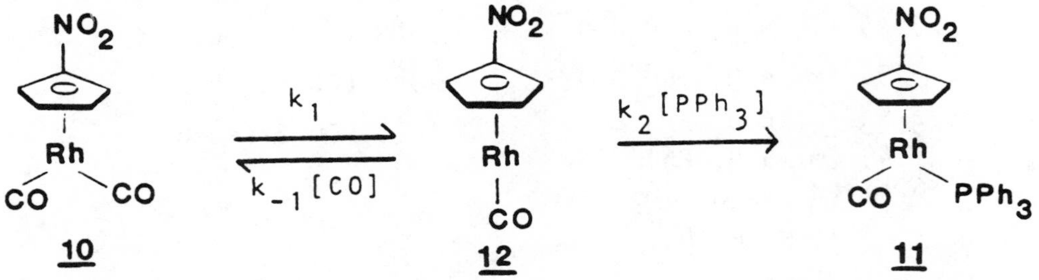

Using a steady state approximation for the 16-electron intermediate **12**, the rate expression derived is:

$$rate = k_{obs}[10], \text{ where}$$

$$k_{obs} = \frac{k_1 k_2 [PPh_3]}{k_{-1}[CO] + k_2[PPh_3]}$$

If $k_2[PPh_3] >> k_{-1}[CO]$, then $k_{obs} = k_1$. This expression does not contain $[PPh_3]$ so it cannot be correct.

If $k_2[PPh_3] << k_{-1}[CO]$, then

$$k_{obs} = \frac{k_1 k_2 [PPh_3]}{k_{-1}[CO]}$$

This expression shows a first order dependence on $[PPh_3]$ but k_{obs} will not be a constant; instead, it will decrease as the reaction proceeds because $[CO]$ is increasing (i.e. CO depresses the rate).

If an intermediate situation exists, the data suffer from both problems, i.e., an inverse dependence on $[CO]$ and no direct dependence on $[PPh_3]$. In addition, a $\Delta S^{\ddagger}$ value that is less than zero is inconsistent with a dissociative mechanism.

2) Associative Mechanism

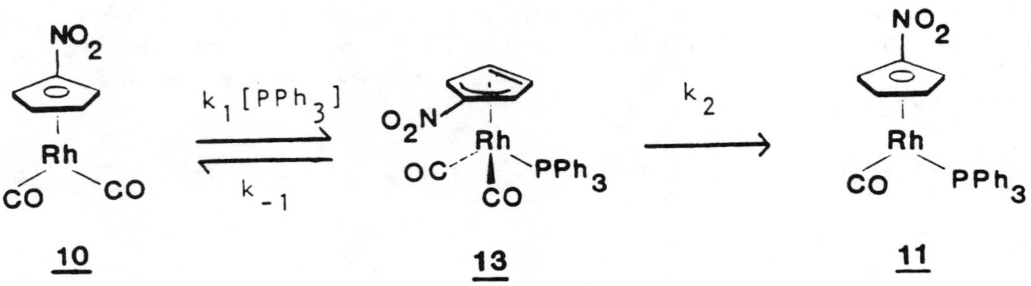

6. (cont.) Slipping of the Cp ring from η^5 to η^3 occurs in order to prevent formation of a 20-electron complex. Using a steady state approximation for the intermediate **13**, the rate expression derived is

$$\text{rate} = k_{obs}[\mathbf{10}], \text{ where}$$

$$k_{obs} = \frac{k_1 k_2 [PPh_3]}{k_{-1} + k_2}$$

This expression gives a first-order dependence on **10** and $[PPh_3]$, with k_{obs} varying linearly with $[PPh_3]$. As for any associative mechanism, $\Delta S^{\ddagger}$ should be less than zero. Therefore, the best mechanism appears to be the associative pathway.

Rerek, M.E.; Basolo, F. J. Am. Chem. Soc. **1984**, 106, 5908.

7. Mechanism I involves CO dissociation from the dimer, with the metal-metal bond remaining intact throughout the reaction. Mechanism II involves initial cleavage of the Re-Re bond, followed by ligand substitution on the 17-electron species $Re(CO)_5$ and recombination of the fragments to form the product dimer. The fact that there are two isotopes of rhenium suggests a double-labeling experiment to distinguish between the two mechanisms. The reaction can be carried out using a 1:1 mixture of $^{185}Re_2(CO)_{10}$ and $^{187}Re_2(CO)_{10}$. If mechanism I were operative, the substitution product would contain only $^{185}Re-^{185}Re$ and $^{187}Re-^{187}Re$ metal-metal bonds in a 1:1 ratio. If mechanism II were operative, the product would contain $^{185}Re-^{185}Re$, $^{185}Re-^{187}Re$, and $^{187}Re-^{187}Re$ metal-metal bonds in a 1:2:1 ratio. These isotope ratios are easily detectable by mass spectroscopic analysis. Experimentally, the thermal reaction proceeds via mechanism I, and the photochemical reaction by mechanism II.

Stolzenberg, A.M.; Muetterties, E.L. J. Am. Chem. Soc. **1983**, 105, 822.

8.   **10** $[CpRe(CO)_2(NO)]^+ BF_4^-$

      11 $CpRe(CO)_2(PPh_3)$

      12 $[CpRe(CO)(NO)(PPh_3)]^+ BF_4^-$

      13 $[CpRe(CO)(NO)(CH_3CN)]^+ BF_4^-$

      14 $[CpRe(PPh_3)_2(NO)]^+ BF_4^-$

8. (cont.)

$$
\begin{array}{c} \\ O \\ \| \end{array}
$$

15   Cp(NO)(PPh$_3$)Re—C—H

16   Cp(NO)(PPh$_3$)Re—CH$_2$OCH$_3$

Points to note about these syntheses:

a)   Both   iodosobenzene   and   trimethylamine—N—oxide   act   as "CO—abstractors" by oxidizing CO to CO$_2$; however, it is evident from this work that iodosobenzene is a more selective CO oxidant than Me$_3$NO.

b) Phosphine substitution will not occur on the nitrosyl complexes, so the phosphine ligand must be introduced before the nitrosyl group.
c) Me$_3$NO does not decompose the phosphine—substituted complexes.

Tam, W.; Lin, G.-Y.; Wong, W.-K.; Kiel, W.A.; Wong, V.K.; Gladysz, J.A. J. Am. Chem. Soc. **1982**, <u>104</u>, 141.

9.   a)

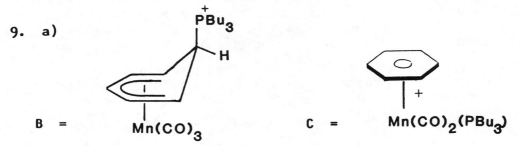

B =   Mn(CO)$_3$ C =   Mn(CO)$_2$(PBu$_3$)

b)

$$ A \quad + \quad L \quad \underset{k_{-1}}{\overset{k_1}{\rightleftharpoons}} \quad B \tag{1} $$

The rate expression for this equation is

$$ -d[A]/dt = k_1[A][L] - k_{-1}[B] \tag{2} $$

which under pseudo-first-order conditions (high [L]), becomes

$$ -d[A]/dt = k_L[A] - k_{-1}[B] \tag{3} $$

where k_L is defined as $k_1[L]$.

In order to get this rate expression into an easily- integrable form, we must consider the following two equalities to find an expression for [B].

1. $[A]_0 + [B]_0 = [A]_\infty + [B]_\infty = [A] + [B]$ (4)

9. (cont.) and since the reaction is an equilibrium,

2. $k_L[A]_\infty = k_{-1}[B]_\infty$ (5)

From equation (4), we define [B],

$$[B] = [A]_\infty + [B]_\infty - [A] \tag{6}$$

and from equation (5), we can make an expression for $[B]_\infty$ in terms of $[A]_\infty$

$$[B]_\infty = (k_L/k_{-1})[A]_\infty \tag{7}$$

and substitute it into equation (6) to give equation (8).

$$[B] = [A]_\infty + (k_L/k_{-1})([A]_\infty) - [A] \tag{8}$$

Substituting this expression for [B] back into equation (3), we obtain equation (9),

$$-d[A]/dt = k_L[A] - k_{-1}\{[A]_\infty + (k_L/k_{-1})([A]_\infty) - [A]\} \tag{9}$$

which after simplification, yields the desired expression, equation (10).

$$-d[A]/dt = (k_1[L] + k_{-1})([A] - [A]_\infty) \tag{10}$$

To integrate, rearrangement is necessary.

$$\int_0^t \frac{d[A]}{[A] - [A]_\infty} = -\int_0^t (k_1[L] + k_{-1})\, dt \tag{11}$$

Integration then yields:

$$\ln \frac{[A]_t - [A]_\infty}{[A]_0 - [A]_\infty} = -(k_1[L] + k_{-1})t \tag{12}$$

and the expression that was plotted is

$$\ln([A]_t - [A]_\infty) = -(k_1[L] + k_{-1})t + \ln([A]_0 - [A]_\infty) \tag{13}$$

c) We know that $K_{eq} = k_1/k_{-1}$ and   (14)

$$\Delta G^\circ = -RT \ln K_{eq} \tag{15}$$

$$\ln K_{eq} = -\Delta G^\circ/RT \tag{16}$$

$$\ln K_{eq} = (-\Delta H^\circ/R)(1/T) + \Delta S^\circ/R \tag{17}$$

so a plot of $\ln K_{eq}$ versus $1/T$ will give a straight line with slope equal to $-\Delta H^\circ/R$ and y-intercept equal to $\Delta S^\circ/R$. Thus:

9. (cont.)

$$-\Delta H^o/R = 5963 \ K^{-1} \quad \text{and} \quad \Delta H^o = 11.8 \text{ kcal/mole}$$

$$\Delta S^o/R = -14.4 \quad \text{and} \quad \Delta S^o = -28.6 \text{ eu}$$

and at 25 ^{o}C, $K_{eq} = 297$.

Kane-Maguire, L.A.P.; Sweigart, D.A. Inorg. Chem. **1979**, <u>18</u>, 700.

For an analysis of the kinetics, see:
Espenson, J.H. "Chemical Kinetics and Reaction Mechanisms" (McGraw-Hill Book Company, 1981), p. 42.

10. The proposed mechanism is shown below.

$$\cdot Mn(CO)_3L_2 \ + \ CO \ \xrightarrow{\ k \ } \ [\cdot Mn(CO)_4L_2]^{\ddagger} \ \longrightarrow \ \cdot Mn(CO)_4L \ + \ L$$

$$2\cdot Mn(CO)_4L \ \xrightarrow{\ \text{fast} \ } \ Mn_2(CO)_8L_2$$

The lack of inhibition of the reaction rate by excess L implies that ligand dissociation is not occurring prior to the rate-determining step. Because the presence of CO <u>increases</u> the rate of the reaction, the initial associative step must be rate-determining. The rate expression is:

$$-d[\cdot Mn(CO)_3L_2]/dt = k[\cdot Mn(CO)_3L_2][CO]$$

An associative mechanism also seems likely in view of the increase in reaction rate upon substitution of the less-bulky phosphine $P(n-bu)_3$ for $P(i-bu)_3$.

The 19-electron complex [$\cdot Mn(CO)_4L_2$] is proposed as the transition state, similar to the transition state for an S_N2 organic substitution reaction.

McCullen, S.B.; Walker, H.W.; Brown, T.L. J. Am. Chem. Soc. **1982**, <u>104</u>, 4007.

Kidd, D.R.; Brown, T.L. J. Am. Chem. Soc. **1978**, <u>100</u>, 4095.

Fawcett, J.P.; Jackson, R.A.; Pöe, A.J. J. Chem. Soc., Dalt. Trans. **1978**, 789.

11. Because carbon is small.

Dewar, M.J.S.; Healy, E. Organomet. **1982**, <u>1</u>, 1705.

4

Rearrangements and Fluxional Processes

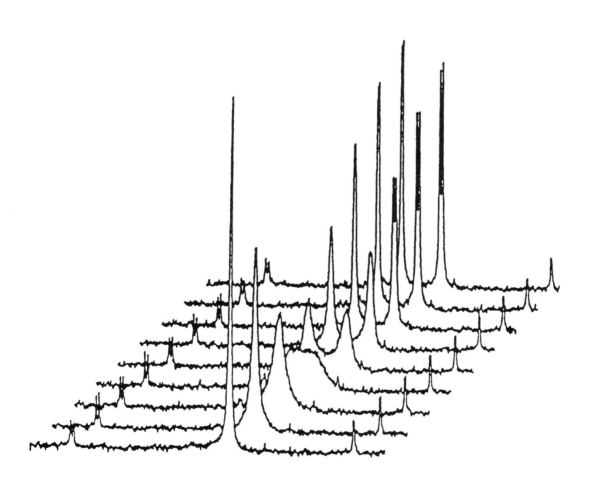

QUESTIONS

Reviewing the concepts covered in Chapter 1 will be helpful for solving some of the problems in this chapter.

1. In the 25 MHz ^{13}C NMR spectrum of a metal complex at $-97^{\circ}C$, there are two signals, one at 182 ppm and one at 178 ppm. At $-58^{\circ}C$ the signals have just coalesced into a broad signal at 180 ppm. What is the rate constant for exchange of the two carbon atoms that produce these signals.

2. Shown below is a section of the variable temperature 250 MHz proton NMR spectrum of $[Cp(CO)_2(PEt_3)W=CH_2]^+$.

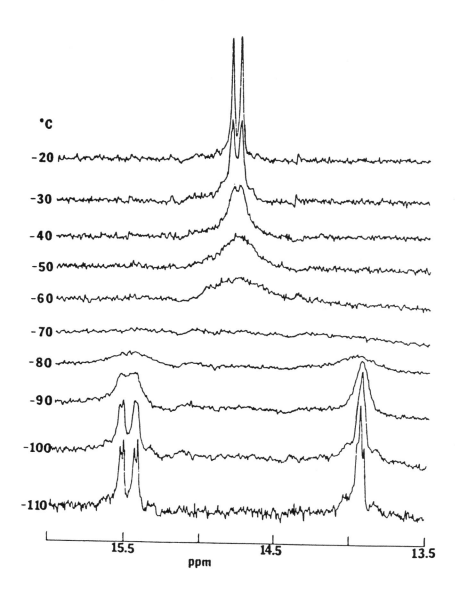

2. (cont.)
a) Suggest a dynamic process that would cause the observed change in the spectrum with temperature.

b) Explain why a doublet is observed at $-20°C$.

c) Explain why two different sets of peaks are observed at $-110°C$.
d) What is the coalescence temperature?
e) What is the rate constant for the dynamic process at the coalescence temperature?
f) What is the free energy of activation, $\Delta G^{\ddagger}$, for this dynamic process?

3. The proton NMR spectrum of $Cp^{*}Rh(C_6(COOCH_3)_6)$ shows the following resonances at $30°C$.

$$\delta\ 1.49\ (s,\ 15\ H)$$
$$\delta\ 3.40\ (s,\ 6\ H)$$
$$\delta\ 3.59\ (s,\ 6\ H)$$
$$\delta\ 3.61\ (s,\ 6\ H)$$

Upon heating the sample, the three closely-spaced peaks coalesce to one singlet while the high-field signal remains unchanged. The signal at $\delta\ 3.59$ is observed to broaden much more rapidly than the signals at $\delta\ 3.40$ and $\delta 3.61$.

a) What is the structure of the compound?

b) Discuss the various processes which could cause the temperature-dependent behavior.

c) What does the observation of selective broadening imply about the assignment of the $\delta 3.59$ signal and about the actual dynamic process occurring?

4. An isomer of $Cp_3Rh_3(CO)_3$ is shown below.

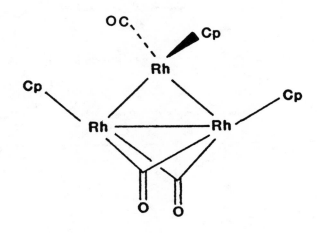

4. (cont.) This isomer is a fluxional molecule, i.e., the carbonyls scramble among the three rhodium atoms. Show the mechanism of the carbonyl scrambling process. In addition, sketch the rate-limiting ^{13}C spectra, i.e., what does the carbonyl region of the NMR spectrum look like for both the low temperature static structure and the high temperature averaged structure. Be sure to indicate approximate chemical shifts for each spectrum.

Important information: Rh has I = 1/2. Approximate values for Rh-C coupling and chemical shifts of carbon bound to rhodium are shown below.

	terminal	μ_2	μ_3
J_{Rh-C}	80	40	30
δ_C	180	200	230

5. The double reduction of benzenemanganese tricarbonyl cation by a powerful hydride source yields one of the three isomeric complexes shown below. (The dotted lines in **B** represent a three-center (Mn-C-H), two-electron bond which is referred to as an agostic (or bridging) hydrogen.)

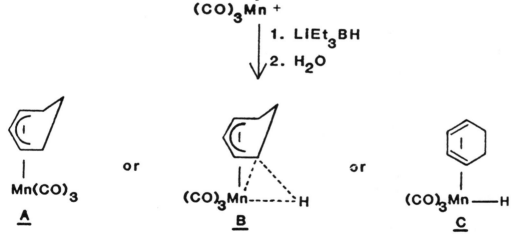

The ground state molecule undergoes <u>two</u> dynamic processes and each of the other two isomers is proposed as an intermediate or transition state for these processes. From the variable temperature 100 MHz NMR spectra given on the following pages:

a) Determine which of three structures is the lowest energy form and explain your reasoning.
b) Propose a mechanism for each of the two dynamic processes.
c) Determine $\Delta G^{\ddagger}$ for these two dynamic processes.

HINT
For the ground state molecule, J_{C-H} for the upfield hydrogen is 85 Hz.

100 MHz ^{1}H NMR spectra of $C_6H_9Mn(CO)_3$

from $-99°$ to $-9°$ C in CD_2Cl_2

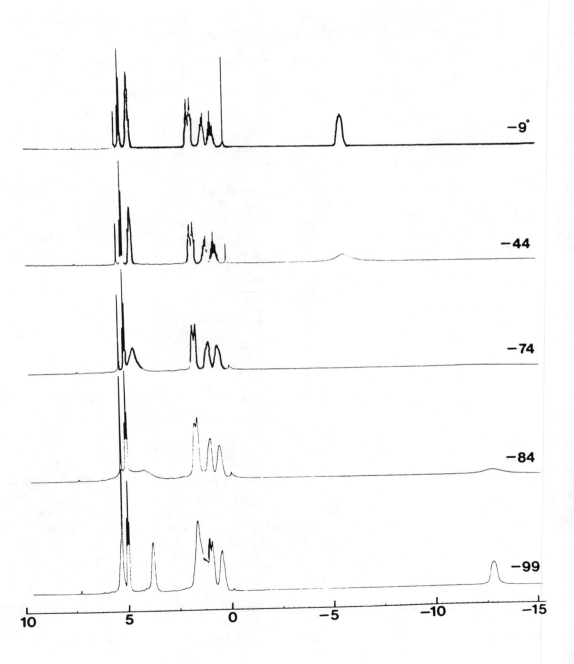

100 MHz ^{1}H NMR spectra of $C_6H_9Mn(CO)_3$

from 4° to 119°C in toluene-d_8

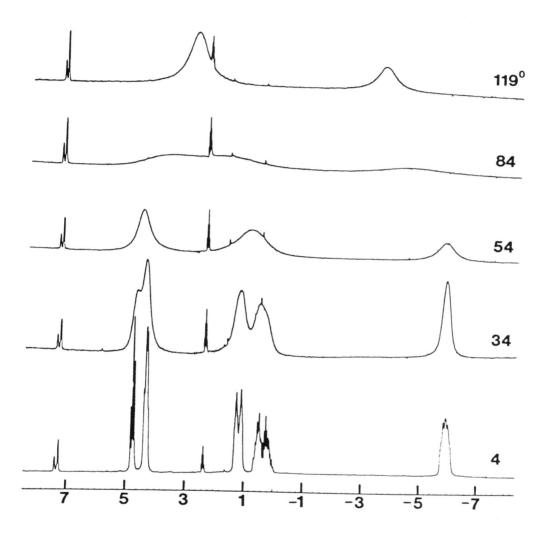

6. The following isomerization reaction (L is an optically-active phosphite) was found to give cyclooctatetraene and optically active vinyl cyclohexene with enantiomeric excesses ranging from 1–20%, depending on the specific phosphite used. Explain how each product is formed. Be sure to rationalize the observed optical activity.

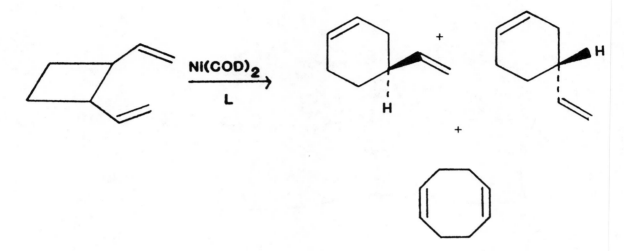

7. Propose a mechanism for the reaction shown below that is consistent with the following experimental observations:

i. The starting material shown (<u>trans</u>-diester) gives exclusively the Z-diene complex, as shown.

ii. The corresponding <u>cis</u>-diester gives exclusively the E-diene complex.

iii. The reaction as shown is very slow; however, if $Fe_2(CO)_9$ is added to the reaction mixture, the product is formed rapidly.

8. Provide a mechanism which explains the following experimental observation.

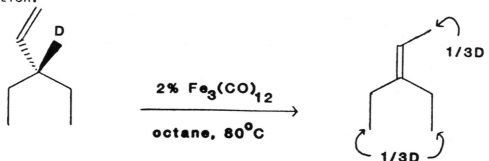

9. Give structures for compounds **D** and **E** and assign the peaks in the NMR spectra given below. Compound **E** is an isomer of **D**.

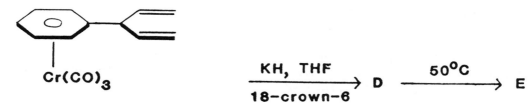

HINT
KH is a non-nucleophilic base.

^{1}H NMR data (δ, THF-d$_{8}$)

D		E	
4.32 (2 H, m)		0.01 (2 H, d)	
4.46 (1 H, m)		2.20 (2 H, d)	
4.60 (2 H, m)		5.13 (2 H, t)	
4.65 (2 H, m)		6.99 (1 H, m)	
5.06 (2 H, m)		7.12 (2 H, m)	
6.35 (2 H, dd)		7.55 (2 H, m)	

10. Consider the following thermal and Pd(II)-catalyzed Cope rearrangements:

![chemical structures 1, 2, 3]

1   **2**   +   **3**

10. (cont.)

Conditions	Yield	Product Ratio (2/3)
240°C	50%	1:1
room temperature, $PdCl_2(CH_3CN)_2$	86%	7:3

In strictly organic systems, the thermal Cope rearrangement proceeds through a "chair" transition state. Propose a mechanism for the Pd(II)-catalyzed reaction.

11. Remembering the "cyclization" mechanism of the Pd(II)-catalyzed Cope rearrangement (see preceeding problem in this chapter), predict the product for the following reaction.

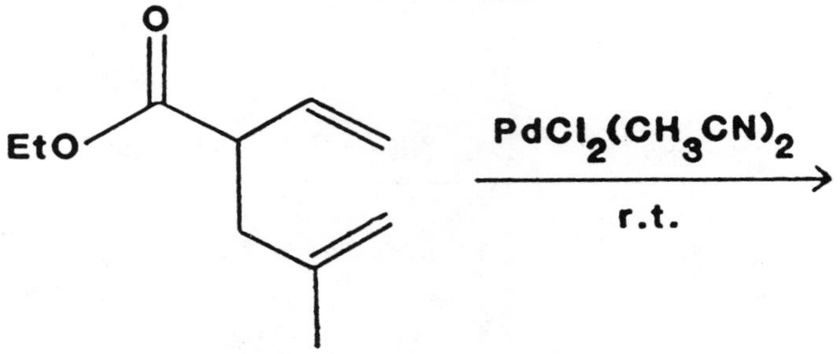

4

Be careful. The product shows only one vinylic resonance in the proton NMR spectrum and an unconjugated ester carbonyl stretch in the IR spectrum.

12. Again, remember the "cyclization" mechanism of the Pd(II)-catalyzed Cope rearrangement (Problem 10, this chapter) and propose a mechanism for the following reaction.

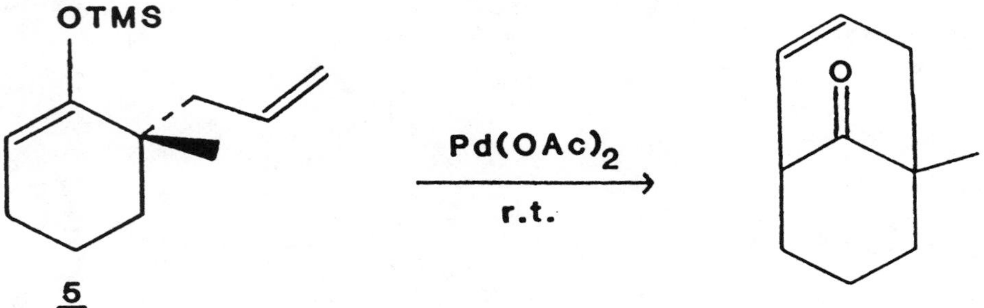

5   **6**

ANSWERS

Reviewing the concepts covered in Chapter 1 will be helpful for solving the problems in this chapter.

1. At the coalescence temperature, the equation for the rate constant is

$$k = \frac{\pi(\Delta\nu)}{\sqrt{2}}$$

The difference in chemical shift of the two signals is 4 ppm, but the formula requires that this shift difference be in Hz. This value depends on the strength of the magnet used for the measurement and can be obtained by multiplying the operating frequency of the magnet (in MHz) and the difference in chemical shifts in ppm. Thus, for this example, we obtain a value of:

$$\Delta\nu = 4 \text{ ppm } (25 \text{ MHz}) = 100 \text{ Hz}$$

and the value of the rate constant at $-58^{\circ}C$ is calculated to be

$$k = (\pi/\sqrt{2})(100 \text{ s}^{-1})$$

$$k = 2.2 \times 10^2 \text{s}^{-1}$$

If a magnet of different field strength were used for this same measurement, the coalescence temperature would also be different.

2. **a)** The dynamic process that averages H_a and H_b is rotation around the W=C bond. Since the two protons are nonequivalent in the static structure, the preferred orientation of the carbene ligand must be the upright position. It is interesting to note that the value of J_{P-Ha} is smaller than J_{P-Hb}.

b) At $-20^{\circ}C$, rotation around the W=C bond is rapid on the NMR timescale and the observed chemical shift is the average of the chemical shifts of H_a and H_b in the static structure. The doublet is due to phosphorus coupling, with the value of the coupling constant equal to $1/2(J_{P-Ha} + J_{P-Hb})$.

2. **c)** At −110°C, the rotation around the W=C bond is slow on the NMR timescale and H_a and H_b appear as non-equivalent hydrogens. In addition, H_a and H_b are coupled to each other and are coupled to phosphorus with $J_{P-Ha} \neq J_{P-Hb}$. The result is two sets of doublet of doublets.

d) $T_c = -70°C$

e) The rate constant for exchange at the coalescence temperature is given by the equation:

$$k = \frac{\pi(\Delta\nu)}{\sqrt{2}}$$

The two resonances are separated by 1.6 ppm and at 250 MHz

$$\Delta\nu = 1.6 \text{ ppm (250 MHz)} = 400 \text{ Hz}$$

and thus, at the coalescence temperature (−70°C)

$$k = (\pi/\sqrt{2})(400 \text{ s}^{-1}) = 888 \text{ s}^{-1}$$

f) The Eyring equation gives the relationship of the rate constant to $\Delta G^{\ddagger}$

$$k = (\kappa T/h) \, e^{-\Delta G^{\ddagger}/RT}$$

thus, $\Delta G^{\ddagger} = -RT[\ln(k/T) + \ln(h/\kappa)]$

where $R = 1.987 \times 10^{-3}$ kcal/mol·K

κ = Boltzmann's constant = 1.38054×10^{-16} erg/K

h = Plank's constant = 6.6256×10^{-27} erg·sec

T = temperature in K

thus, for this example

$$\Delta G^{\ddagger} = 9.0 \text{ kcal/mol}$$

Kegley, S.E.; Brookhart, M.; Husk, G.R. Organomet. **1982**, <u>1</u>, 760.

3. **a)** One type of methyl on the Cp ring and three types of carbomethoxy groups on the benzene ring are most easily accounted for by the 18-electron species shown below.

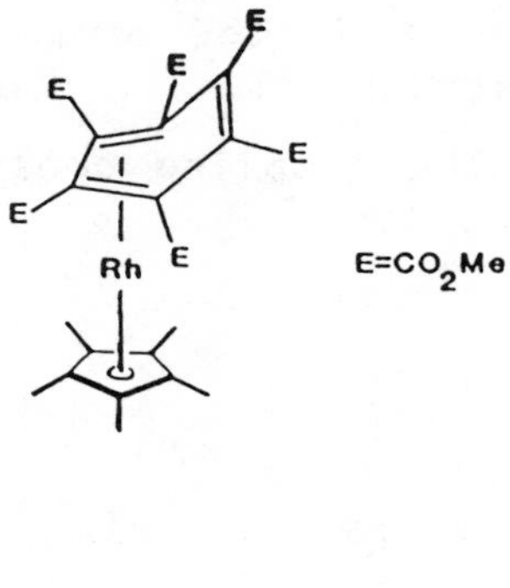

E=CO₂Me

3. (cont.) This compound is isoelectronic with η^6, η^4-$(C_6H_6)_2Ru$. (Darensbourg, M.Y.; Muetterties, E.L. J. Am. Chem. Soc. **1978**, **100**, 7425.)

b) The methyl groups on the Cp* ring remain equivalent at all temperatures, while the carbomethoxy groups exchange. This observation can be explained by the Cp*Rh unit moving around the benzene ring, which can be accomplished in a number of ways: 1) The Cp*Rh can shift one bond at a time around the ring (a 1,2-shift), 2) two bonds at a time (a 1,3-shift), 3) three bonds at a time (a 1,4-shift), or 4) randomly. If the three resonances are assigned a, b, and c as shown below, the results of three of the above processes can be depicted as follows:

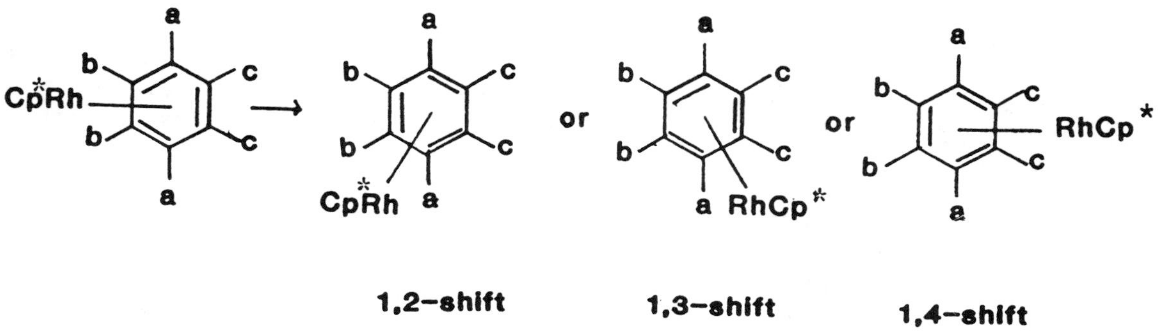

1,2-shift   **1,3-shift**   **1,4-shift**

Redefining each resonance of the products in terms of the type of proton it was in the starting material gives the following information:

1,2-shift	1,3-shift	1,4-shift
a ⟶ c	a ⟶ c	a ⟶ a
b ⟶ a	b ⟶ c	b ⟶ c
b ⟶ b	b ⟶ a	b ⟶ c
a ⟶ b	a ⟶ b	a ⟶ a
c ⟶ a	c ⟶ b	c ⟶ b
c ⟶ c	c ⟶ a	c ⟶ b

c) If the scrambling were occurring via a random process, each signal must broaden equally. Since this is not observed, the random process can be eliminated.

If a 1,4-shift were occurring, the "a" methyl signals can not broaden since they are not exchanging. This is also not observed and the 1,4-shift can be eliminated.

If a 1,3-shift were occurring, all methyls must exchange at an equal rate, thus the rate of broadening must be the same for methyls a, b, and c. This is not observed.

If a 1,2-shift were occurring, the "a" methyl group exchanges twice as fast as "b" and "c"; as a result, the signal due to "a" should broaden twice as fast as the others. This is in fact what is observed and allows assignment of the signal at $\delta 3.59$ to the carbomethoxy groups at position "a". The other two resonances cannot be assigned absolutely.

3. c) (cont.) Additional point to note:

When the fluxional process is very fast on the NMR timescale, a single carbomethoxy signal will appear at a chemical shift that is the average of the three contributing signals, e.g.

$$\delta = \frac{3.40 + 3.59 + 3.61}{3} = 3.53 \text{ ppm}$$

Kang, J.W.; Childs, R.F; Maitlis, P.M. _J_. _Am_. _Chem_. _Soc_. **1970**, _92_, 720.

4. The NMR spectrum of the static structure (Figure 1) should show three types of carbonyls, one terminal and two bridging. Note that the two μ^2-carbonyls are <u>not</u> equivalent—one is <u>cis</u> to a Cp and the other is <u>cis</u> to a CO. The terminal CO should appear as a doublet since it is coupled to one rhodium. The bridging CO's should appear as a triplet since they are coupled to two rhodium atoms.

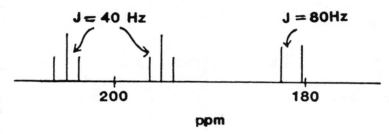

Figure 1: Carbon-13 NMR spectrum of the static structure

The fluxional process is outlined below.

4. (cont.) Notice that by this mechanism, each CO can bond to each Rh atom; however, one CO is unique since it is always on the top face of the Rh–Rh–Rh plane while the other two are on the bottom face (and are equivalent). Thus in a static structure, the CO on top is always bridging, and of the other two, one is bridging and one is terminal.

The NMR spectrum of the fluxional molecule (Figure 2, next page) should show the following:

1) Two sets of peaks in a ratio of 1:2.
2) The unique CO will be at approximately 200 ppm, since it is always μ^2.
3) Each of the two other CO's bridges half of the time and is terminal half of the time; thus the other set of peaks will appear at

$$\delta = \frac{200 \text{ ppm} + 180 \text{ ppm}}{2} = 190 \text{ ppm}$$

4) Each CO is coupled to all three Rh atoms, thus each peak will appear as a 1:2:2:1 quartet.
5) Coupling constants: For both types of CO, we must average all J_{Rh-C}.

For the unique CO:

$$J_{Rh(1)-C} = 40 \text{ Hz}$$
$$J_{Rh(2)-C} = 40 \text{ Hz}$$
$$J_{Rh(3)-C} = 0 \text{ Hz}$$

$$\overline{\phantom{J_{Rh(3)-C} = 0 \text{ Hz}}}$$

$$J_{Rh(avg.)-C} = 27 \text{ Hz}$$

For the equivalent CO's:

$$J_{avg.} = 1/2(\text{bridging } J_{Rh-C} + \text{terminal } J_{Rh-C})$$

Bridging

$$J_{Rh(1)-C} = 40 \text{ Hz}$$
$$J_{Rh(2)-C} = 40 \text{ Hz}$$
$$J_{Rh(3)-C} = 0 \text{ Hz}$$

$$J_{Rh(avg.)-C} = 27 \text{ Hz}$$

Terminal

$$J_{Rh(1)-C} = 0 \text{ Hz}$$
$$J_{Rh(2)-C} = 0 \text{ Hz}$$
$$J_{Rh(3)-C} = 80 \text{ Hz}$$

$$J_{Rh(avg.)-C} = 27 \text{ Hz}$$

Thus, the overall $J_{avg.} = 27$ Hz.

4. (cont.)

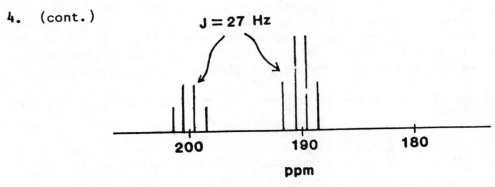

Figure 2: Carbon-13 NMR spectrum of the fluxional molecule

Lawson, R.J.; Shapley, J.R. Inorg. Chem. **1978**, 17, 772.

Lawson, R.J.; Shapley, J.R. J. Am. Chem. Soc. **1976**, 98, 7433.

5. a) The lowest energy form will be that observed as the static structure at the low temperature limit.

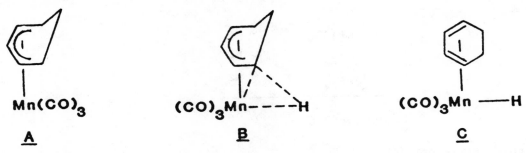

The spectrum obtained at −99°C rules out structure A. Structure A would be expected to have only alkyl and allylic protons and no resonances at −12.8 ppm. Both structures B and C would be expected to have an upfield resonance. Structure C has a mirror plane and thus five different types of protons. On the other hand, B has no mirror plane, and all nine protons are inequivalent. The low temperature spectrum is thus more consistent with B than with C. The observed J_{C-H} of 85 Hz is also indicative of a significant amount of C-H interaction. An approximate value of J_{C-H} can be obtained by the equation

$$J = 500(\rho)$$

where ρ is the percent s-character of the orbitals involved. For structure C, no C-H interaction is involved. For structure B, where the amount of C-H interaction is less than that in a pure sp^3-hybridized system, we would expect a value of J_{C-H} that is somewhat less than 125 Hz.

b) Both dynamic processes disrupt the three-center, two-electron bond. The lower temperature process breaks the metal-hydrogen interaction.

5. b) (cont.)

When the bridge is reformed, either of two <u>endo</u> hydrogens can bond to the metal. Thus, at −9°C, the upfield peak is shifted to −5.7 ppm which is the average of the shifts for the two <u>endo</u> hydrogens (1.4 and −12.8 ppm) in the static structure observed at −99°C. This dynamic process introduces a mirror plane and thus the spectrum simplifies to only six peaks.

As the temperature continues to increase, a second dynamic process which breaks the carbon–hydrogen interaction begins.

When the bridging hydrogen is reformed, the hydrogen can go to either side of the diene. This process, in conjunction with the lower-temperature process (which is quite rapid since the temperature is well above −9°C), will make all three <u>endo</u> hydrogens equivalent. The result is that at 119°C, only two peaks are observed.

c) Spin saturation transfer experiments (SST) were used to accurately determine the values of $\Delta G^{\ddagger}$ for these two processes; however, approximate values can be obtained using line broadening.

For the low-temperature process:

$$T_c = -74°C$$

and the rate constant for exchange at the coalescence temperature can be calculated by the following formula.

$$k = \frac{\pi(\nu_a - \nu_x)}{\sqrt{2}} = \frac{\pi[140-(-1280)Hz]}{\sqrt{2}} = 3154 \text{ s}^{-1}$$

5. c) (cont.) The Eyring equation gives the relationship of the rate constant to $\Delta G^{\ddagger}$

$$k = (\kappa T/h)\, e^{-\Delta G^{\ddagger}/RT}$$

thus, $\Delta G^{\ddagger} = -RT[\ln(k/T) + \ln(h/\kappa)]$

where $R = 1.987 \times 10^{-3}$ kcal/mol·K

κ = Boltzmann's constant = 1.38054×10^{-16} erg/K

h = Plank's constant = 6.6256×10^{-27} erg·sec

T = temperature in K

and for the low temperature process, $\Delta G^{\ddagger} = 8.3$ kcal/mol. For the high-temperature process, the calculation is repeated, considering the two-proton resonance at −5.7 and the one-proton resonance at 0.5 ppm.

$$T_c = 84^{\circ}C$$

$$k = \frac{\pi(\nu_a - 2\nu_x)}{\sqrt{2}} = \frac{\pi[50-2(-570\ \text{Hz})]}{\sqrt{2}} = 2644\ s^{-1}$$

$$\Delta G^{\ddagger} = 15.5\ \text{kcal/mol}$$

These values agree fairly well with the more accurate values determined by SST: $\Delta G^{\ddagger}$ (low-temperature process) = 8.3 kcal/mol and $\Delta G^{\ddagger}$ (high-temperature process) = 15.4 kcal/mol.

Brookhart, M.; Lamanna, W.; Humphrey, M.B. _J. Am. Chem. Soc._ **1982**, 104, 2117.

6. The reaction proceeds via an insertion of Ni(0) into the C—C bond of cyclobutane.

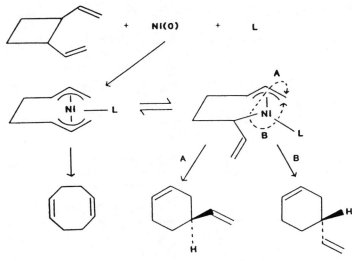

6. (cont.) Because L is optically active, the transition state for ring closure to cyclohexene via pathway A is different in energy than the transition state for ring closure via pathway B. Thus, closure to the cyclohexene products via the different pathways will produce compounds of opposite chirality. Because cyclooctadiene contains a plane of symmetry, no optical activity is observed for this product

Richter, W.J. _J. Mol. Cat._ **1983,** <u>18</u>, 145.

7. Two different mechanisms for this reaction have been proposed in the literature.

<u>Mechanism I</u>

$$Fe_2(CO)_9 \longrightarrow Fe(CO)_4 + Fe(CO)_5$$

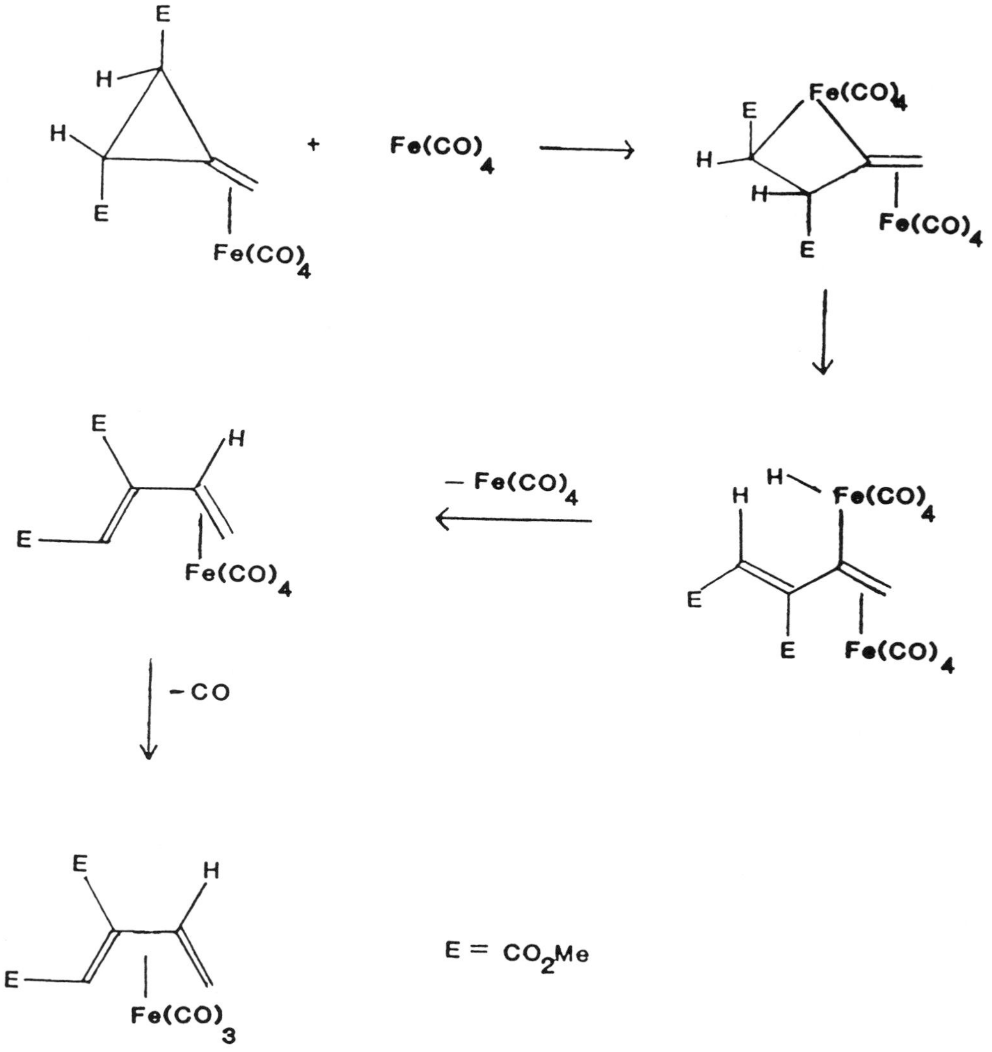

E = CO_2Me

Whitesides, T.H.; Slaven, R.W. _J. Organomet. Chem._ **1974,** <u>67</u>, 99.

7. (cont.) <u>Mechanism II</u>

$$Fe_2(CO)_9 \longrightarrow Fe(CO)_4 + Fe(CO)_5$$

disrotatory opening
of the 2-3 bond

$E = CO_2Me$

Pinhas, A.R.; Samuelson, A.G.; Risemberg, R.; Arnold, E.V.; Clardy, J.; Carpenter, B.K. J. Am. Chem. Soc. **1981**, <u>103</u>, 1668.

For both of these mechanisms, addition of $Fe_2(CO)_9$ should increase the rate of the reaction by providing a source of $Fe(CO)_4$. In the absence of added $Fe_2(CO)_9$, the starting olefin–$Fe(CO)_4$ complex is the source of $Fe(CO)_4$.

8. The postulated mechanism is shown below.

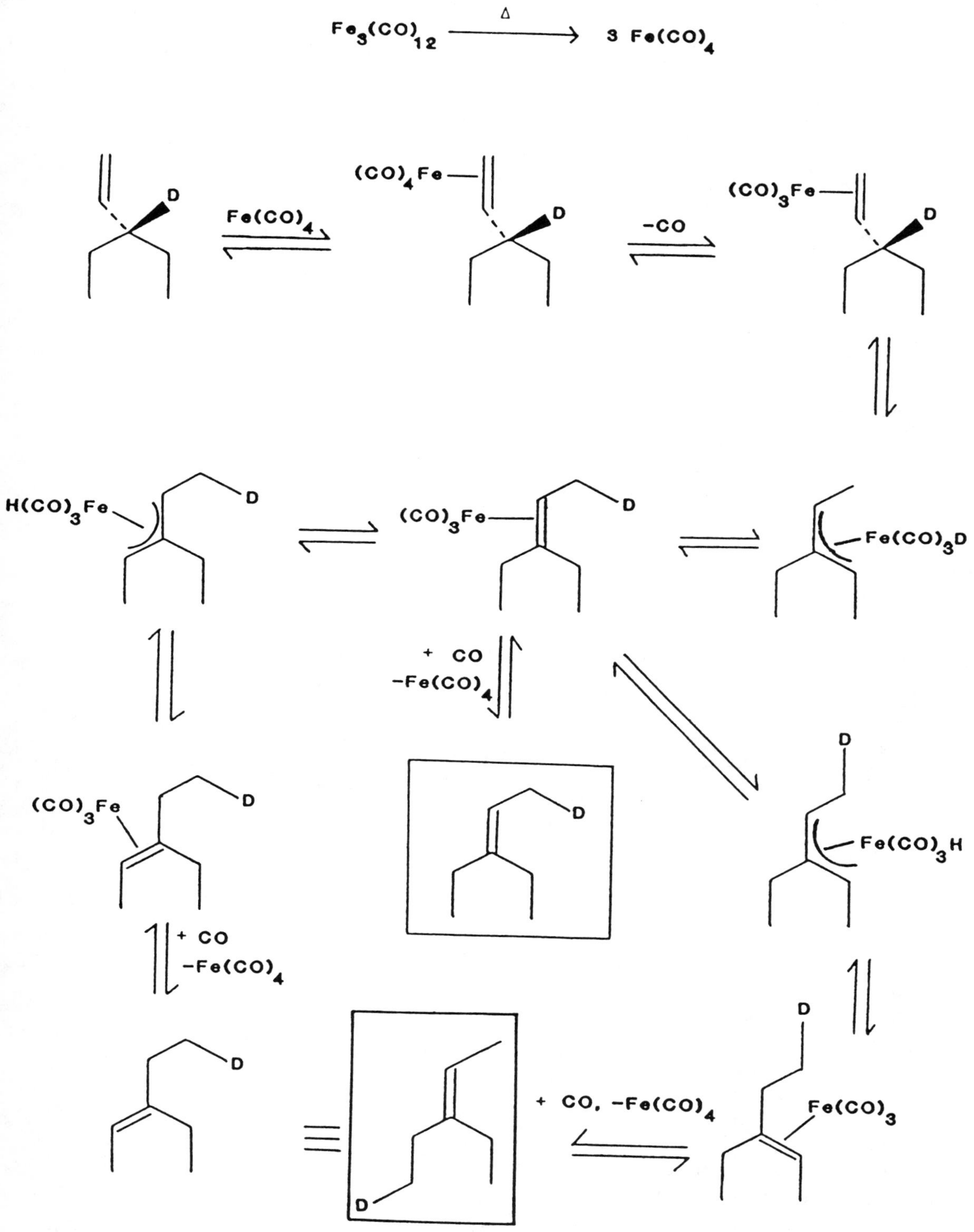

$$Fe_3(CO)_{12} \xrightarrow{\Delta} 3\ Fe(CO)_4$$

8. (cont.)

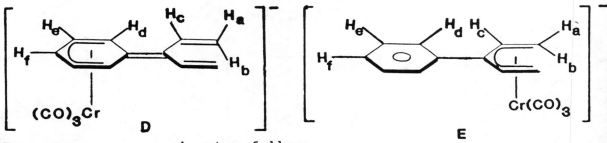

Casey, C.P.; Cyr, C.R. J. Am. Chem. Soc. **1973**, 95, 2248.

9. The products **D** and **E** are shown below.

The resonances are assigned as follows.

D   δ 4.32 (2 H, m, a)
        δ 4.46 (1 H, m, f)
        δ 4.60 (2 H, m, d)
        δ 4.65 (2 H, m, b)
        δ 5.06 (2 H, m, e)
        δ 6.35 (2 H, dd, c)

E   δ 0.01 (2 H, d, b)
        δ 2.20 (2 H, d, a)
        δ 5.13 (2 H, t, c)
        δ 6.99 (1 H, m, f)
        δ 7.12 (2 H, m, e)
        δ 7.55 (2 H, m, d)

The peak at δ 5.13 in the spectrum of **E** should be a doublet of doublets; however, since the two coupling constants are equivalent, the resonance appears as a triplet.

Ceccon, A.; Gambaro, A.; Venzo, A. J. Chem. Soc., Chem. Comm. **1985**, 540.

10. Since the Pd(II)-catalyzed reaction gives the same products as the thermal rearrangement (although in a different ratio), it is reasonable to assume that this reaction also proceeds via an intermediate with a chair conformation. Two different mechanisms have been proposed.

Mechanism I

10. (cont.) <u>Mechanism II</u>

10. (cont.) In mechanism I, a bis-π-allyl Pd(IV) complex is formed. Intermediate **X** has the "chair-like" conformation required to give the observed product, i.e. the open sides of the allyl moieties are pointing in opposite directions. Intermediate **Y** has a "boat-like" conformation, and since product **Z** is not observed, we can assume either that **Y** is not present or that no product formation occurs via this intermediate.

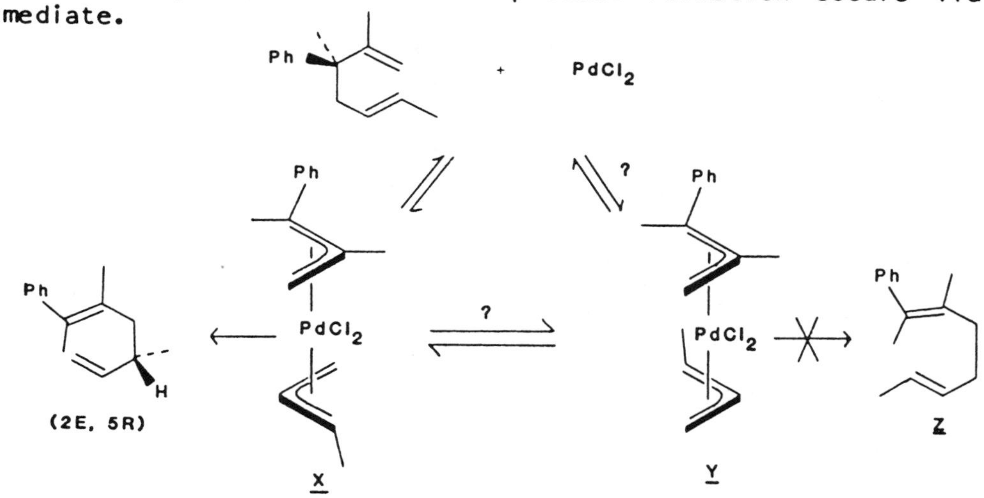

(2E, 5R)

X   Y   Z

In mechanism II, a cyclohexyl-palladium intermediate is formed. Due to the large number of substituents on the ring, the chair conformation is highly favored over the boat. In this mechanism, the cyclization step is a nucleophilic attack on a Pd(II)-coordinated double bond in which the nucleophile is also an olefin. The intermediate can proceed to the product by breaking the C_1-C_2 bond or back to the starting material by breaking the C_4-C_5 bond.

At the present time, neither mechanism has been disproven and each is favored by a different research group.

Henry, P.M. _J._ _Am._ _Chem._ _Soc._ **1972**, _94_, 5200.

Hamilton, R.; Mitchell, T.R.B.; Rooney, J.J. _J._ _Chem._ _Soc._, _Chem._ _Comm._ **1981**, 456.

Overman, L.E.; Jacobsen, E.J. _J._ _Am._ _Chem._ _Soc._ **1982**, _104_, 7225.

Overman, L.E. _Angew._ _Chem._ _Int._ _Ed._ _Engl._ **1984**, _23_, 579.

11. The observed product is shown below.

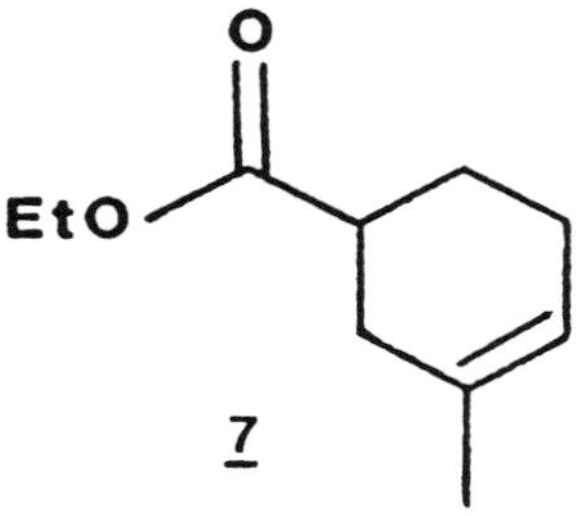

_7_

11. (cont.) The proposed mechanism is:

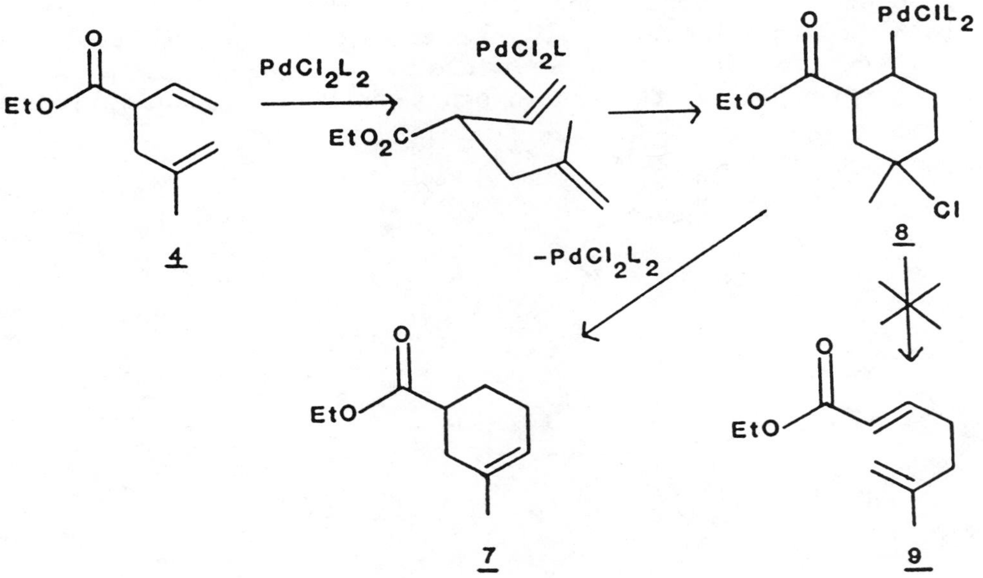

The cyclic intermediate <u>8</u> (formed in the same manner as in Mechanism II, problem **10**) does <u>not</u> give the Cope product due to the instability of the palladium–olefin complex <u>9</u> formed from an electron–deficient olefin. Instead, <u>8</u> gives the cyclic product <u>7</u> by an unknown mechanism. A possible (but unproven) mechanism is shown below.

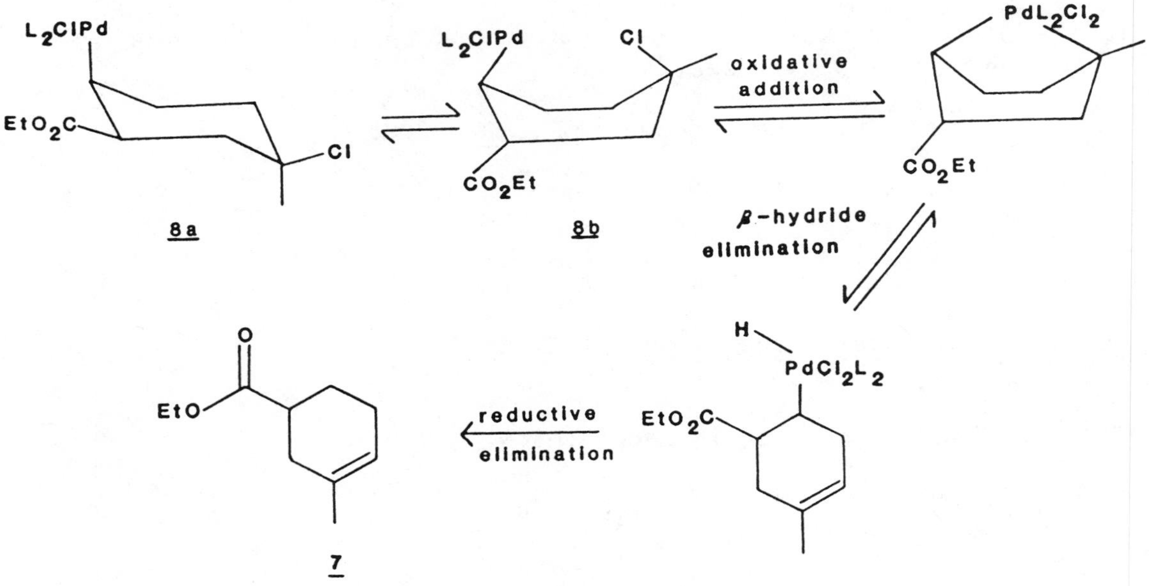

Overman, L.E.; Renaldo, A.F. <u>Tet</u>. <u>Let</u>. **1983**,<u>24</u>, 2235.

12. As in problem **10**, the first step is a nucleophilic attack by one olefin on the coordinated olefin.

It is not clear why this reaction does not give the Cope rearrangement product.

Kende, A.S.; Roth, B.; Sanfilippo, P.J. _J_. _Am_. _Chem_. _Soc_., **1982**, _104_, 1784.

Kende, A.S.; Roth, B.; Sanfilippo, P.J.; Blacklock, T.J. _ibid_, 5808.

Oxidative-Addition and Reductive-Elimination Reactions

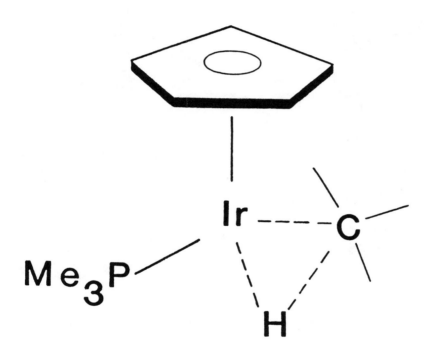

QUESTIONS

1. Predict the products of the following reactions and classify them as either oxidative-addition or reductive-elimination. Indicate the change that occurs in the formal oxidation state of the metal.

a) $CpRh(CO)_2$ + CF_3CF_2I $\xrightarrow{30°C}$

b) $Ir(PPh_3)_3Cl$ $\xrightarrow{heat}$

c) $Cp^*Ir(CO)_2$ $\xrightarrow{h\nu,\ neopentane}$

d) $Cp_2W(CH_3)H$ $\xrightarrow{\Delta,\ benzene}$

e) $Ir(PPh_2Me)_2(CO)Cl$ + CF_3I $\xrightarrow{25°C,\ 3\ min.}$

f) $(diphos)Ir(CO)Br$ + Et_3Si-H $\longrightarrow$ 4 possible products

g) $Cp(CO)_2Fe-CH_3$ + HBF_4 $\longrightarrow$

h) $Cp(CO)_3Re$ + Br_2 $\longrightarrow$

i) $Cp^*(PMe_3)IrH_2$ + n-pentane $\xrightarrow{h\nu}$ 4 products

2. For each pair of complexes given below, predict which one will be more reactive towards oxidative-addition of H_2.

a) $[Co(dppe)_2]^+$ or $[Ir(dppe)_2]^+$

b) $RhCl(PPh_3)_3$ or $RhCl(CO)(PPh_3)_2$

c) $RhCl(CO)(PPh_3)_2$ or $IrCl(CO)(PPh_3)_2$

d) $Ir(CO)(PPh_3)_2Cl$ or $Ir(CO)(PPh_3)_2Br$

e) $[Rh(dppe)_2]^+$ or $[Rh(dmpe)_2]^+$

f) $Ir(PPh_3)_2(CO)Cl$ or $[Pt(PPh_3)_2(CO)Cl]^+$

g) $Os(CO)_5$ or $trans-(PPh_3)_2Os(CO)_3$

3. The complex $(Ph_3P)_3Rh-CH_3$ is cleaved by D_2 into $(Ph_3P)_2Rh(H)(PPh_2(C_6H_4D))$ and CH_4. Suggest a mechanism for this reaction.

4. The following reaction is observed:

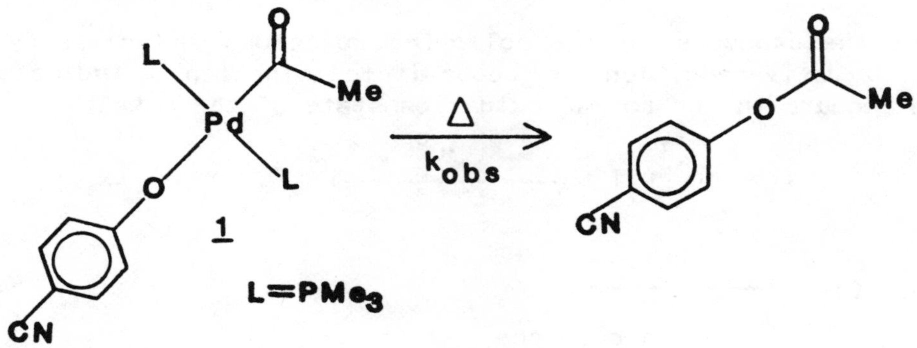

$$L = PMe_3$$

OBSERVATIONS

1) The rate law is of the form $-d[1]/dt = k_{obs}[1]$.

2) Addition of excess PMe_3 decreases the rate of the reaction up to a concentration of 0.4 M. At higher concentrations, no further depression of the rate is noted.

3) In contrast, addition of dimethylphosphinoethane (dmpe) <u>increases</u> the rate of this reaction.

a) Suggest a mechanism for this reductive-elimination reaction.

b) In the context of your proposed mechanism, explain the observed changes in rate with a change in the <u>type</u> of phosphine added.

5. Identify the intermediates 2, 3, 4, 5, and 6 in the following catalytic reaction. Clearly indicate which steps involve oxidative-addition or reductive-elimination.

$$Pd(PPh_3)_4 \quad + \quad Ph-Br \longrightarrow 2$$

NaH   + 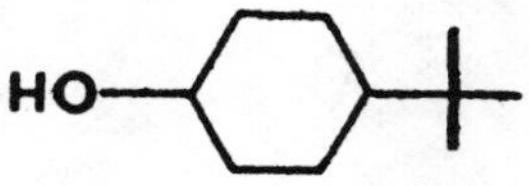   HO——————⟶ 3   no, there's not supposed to be a metal in this step)

$$2 \quad + \quad 3 \longrightarrow 4 \xrightarrow[\text{elimination}]{\beta-\text{hydrogen}} 5$$

5 ⟶ ——=O   +   6

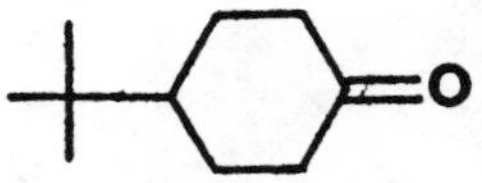

$$6 \longrightarrow Ph-H \quad + \quad Pd(PPh_3)_2$$

6. The following two reactions have been shown to proceed via different mechanisms:

trans-$(PPh_3)_2Ir(CO)Cl$ (**7**)   + H_2 ───────────→ **8**   (1)

7   + $H_3C-O-SO_2CF_3$ ───────────→ **9**   (2)

a) Predict the products, including stereochemistry, and present a reasonable mechanism for each reaction.

b) Speculate as to why different mechanisms prevail for the different reactions.

7. The following reaction is observed:

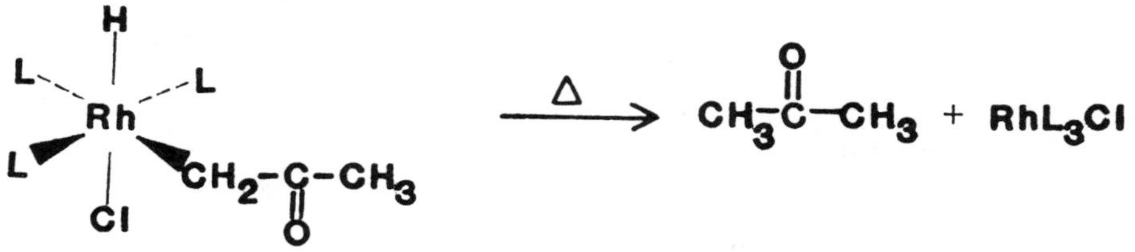

OBSERVATIONS
1) When the reaction is carried out with equimolar amounts of **10**-d_0 and **10**-d_6, only d_0-acetone and d_6-acetone were obtained.

2) The rate of disappearance of **10** follows first order kinetics and is not affected by variations in the initial concentration of **10**.

3) The rate decreases when PMe_3 is added, although first order kinetics are still observed. Additionally, the value of $1/k_{obs}$ varies linearly with the amount of PMe_3 added.

4) The rate increases when "phosphine sponge" ($Rh(acac)(ethylene)_2$) is added.

5) A kinetic isotope effect of k^H_{obs}/k^D_{obs} = 1.3 is observed for the reaction.

7. (cont.) Your mission, should you choose to accept, is to:

a) postulate a mechanism consistent with the observations given above;

b) derive the rate expression for your mechanism.

This page will self-destruct in 10 seconds. Good luck.

8. Oxidation potentials for several ruthenium compounds are shown in the table below.

Compound	$(E_p)_a/V^*$
$CpRu(CO)_2Me$	1.55
$CpRu(CO)(PPh_3)Me$	1.02
$CpRu(CO)(PPh_3)CH_2Ph$	1.03
$CpRu(PPh_3)_2Me$	0.39
$CpRu(PPh_3)_2CH_2Ph$	0.40

$^*(E_p)_a$ = anodic peak potential vs Ag/AgCl/aqueous KCl/reference, 0.1 V/s, CH_2Cl_2 solvent, 0.05 M Bu_4NBF_4 supporting electrolyte.

a) Predict the relative reactivities of the compounds in the table towards electrophilic attack by HgX_2, CuX_2 (X = Cl, Br), and HCl. Explain your reasoning.

b) Predict the products of the following reactions and propose reasonable intermediates.

1)

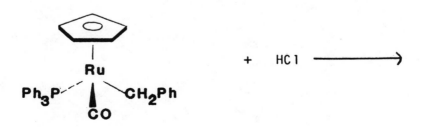

8. (cont.)

2)

3)

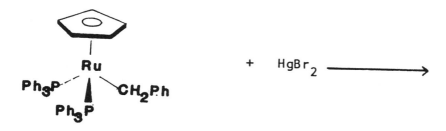

4)

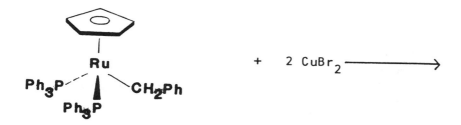

9. The following oxidative-addition reactions have been shown to occur:

9. (cont.) When reaction (2) is carried out, the final products are $Et_3Si-SiEt_3$, $(dppe)Ir(CO)(Br)H_2$, and $(dppe)Ir(CO)(SiEt_3)H_2$. Propose a mechanism that accounts for all products.

10. The reaction of H_2 with $(dppe)Ir(CO)Br$, **11**, in acetone initially gives the kinetic product **12** which isomerizes to the thermodynamic product **13** with $t_{1/2}$ = 35 h at $25^\circ C$. Additionally, mixing **12** with D_2 gives **12-d$_2$** with $t_{1/2}$ = 10 min. at $25^\circ C$.

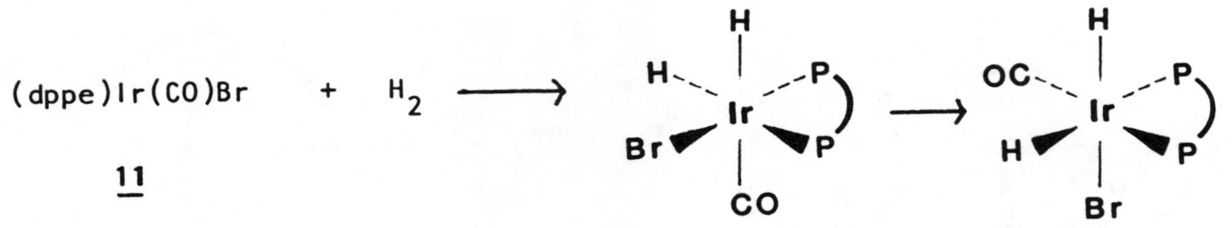

$(dppe)Ir(CO)Br$ + H_2 $\longrightarrow$

11

12   **13**

Two mechanisms proposed for the interconversion of **12** and **13** are:

1) an intramolecular rearrangement of **12**, or

2) reductive-elimination of hydrogen to reform **11**, followed by oxidative addition of hydrogen with a different stereochemistry.

a) Derive a rate expression for mechanism 2), assuming that **11** and **12** are in equilibrium.

b) Derive a rate expression for the intramolecular rearrangement. Are there any differences in the observables between this expression and the one derived for the two-step mechanism?

c) Experimentally, how could the rate of deuterium incorporation be monitored?

11. Products **14** and **15** are obtained from the following reaction.

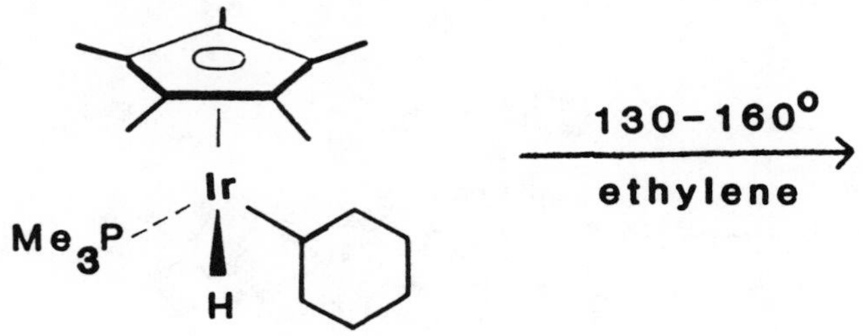

$\xrightarrow[\text{ethylene}]{130-160^\circ}$ **14**   +   **15**

11. (cont.) At higher temperatures, **14** is converted to **15**; however, **15** can not be converted to **14**.

Spectral data for **14**:   Spectral data for **15**:

^{1}H NMR (δ, C_6D_6) ^{1}H NMR (δ, C_6D_6)

-16.88 (d, J=36 Hz, 1 H)   1.86 (d, J=1.5 Hz, 15 H)
1.86 (dd, J=1.8, 0.8 Hz, 15 H)   1.01 (d, J=9.0 Hz, 9 H)
1.24 (d, J=10.1 Hz, 9 H)   1.13, 1.27 (m, 4 H)
5.75, 6.95 (m, 2 H)
8.12 (ddd, J=3.3, 10.2, 17.5 Hz, 1 H)

^{13}C (δ, C_6D_6, broad-band decoupled) ^{13}C (δ, C_6D_6, gated decoupled)

10.27 (s)   10.29 (q, J=126.1)
18.82 (d, J=38.6 Hz)   10.55 (dd, J=150.6, 150.6 Hz)
92.50 (d, J=2.6 Hz)   16.93 (dq, J=33.9, 126.6 Hz)
123.78 (d, J=3.0 Hz)   90.19 (s)
129.18 (d, J=13.4 Hz)

IR (KBr, cm^{-1})   IR (C_6H_6, cm^{-1})

2903, 2105, 1553, 953, 940   2960, 2900, 1360, 940

a) From the data given, identify **14** and **15**. Assign all resonances in the NMR spectra.

b) Postulate a mechanism for the reaction.

ANSWERS

1. a) CpRh(CO)(CF$_2$CF$_3$)I + CO, oxidative-addition, Rh(I)$\longrightarrow$Rh(III)

Collman, J.P.; Roper, W.R. Adv. Organomet. Chem. 1968, 7, 81.

b)

oxidative-addition
Ir(I)$\longrightarrow$Ir(III)

c) The initial product of this reaction is Cp*IrCO, formed by photolytically-induced loss of CO. The sixteen-electron intermediate then undergoes an oxidative-addition reaction with neopentane to form Cp*Ir(CO)(CH$_2$C(CH$_3$)$_3$)H, with a formal oxidation state change from Ir(I) to Ir(III).

Hoyano, J.K.; Graham, W.A.G. J. Am. Chem. Soc. 1982, 104, 3723.

d) The initial product of this reaction is Cp$_2$W, formed by reductive-elimination of methane. The formal oxidation state change for this process is W(IV) to W(II). The sixteen-electron intermediate then undergoes an oxidative-addition reaction with the solvent benzene to form Cp$_2$W(Ph)H, with a formal oxidation state change from W(II) to W(IV).

Cooper, N.J.; Green, M.L.H.; Mahtab, R. J. Chem. Soc., Dalton Trans. 1979, 1557.

e)

oxidative-addition
Ir(I)$\longrightarrow$Ir(III)

Collman, J.P.; Roper, W.R. Adv. Organomet. Chem. 1968, 7, 81.

f) The four possible products resulting from oxidative-addition of the silane are shown on the following page. All oxidation state changes are Ir(I) to Ir(III).

1. f) (cont.)

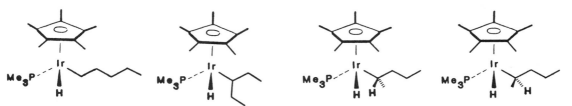

Johnson, C.E.; Eisenberg, R. J. Am. Chem. Soc. 1985, 107, 6531.

g) The initial product of this reaction is the protonated complex $[Cp(CO)_2Fe(CH_3)H]^+$ with a formal oxidation state change of Fe(II) to Fe(IV). The next step is reductive-elimination of CH_4, leaving $[Cp(CO)_2Fe]^+$ $[BF_4]^-$. The formal oxidation state change for this step is Fe(IV) to Fe(II).

h) $Cp(CO)_2ReBr_2$ + CO, oxidative-addition, Re(I)$\longrightarrow$Re(III).

Einstein, F.W.B.; Klahn-Oliva, A.H.; Sutton, D.; Tyers, K.G. Organomet. 1986, 5, 53.

i) The initial product of this reaction is $Cp^*Ir(PMe_3)$, formed by reductive-elimination of H_2. The formal oxidation state change for this process is Ir(III) to Ir(I). The four different products shown below result from oxidative-addition [Ir(I) to Ir(III)] to the different types of C–H bonds in n-pentane. Because the metal center is chiral, addition at the 2-position of n-pentane gives rise to two products that are diastereomeric.

Janowicz, A.H.; Bergman, R.G. J. Am. Chem. Soc. 1983, 105, 3929.

2. In general, the tendency of a complex to undergo oxidative-addition reactions is governed by the amount of electron density at the metal, i.e. by the ease with which the metal can be oxidized. Thus, several general trends have been observed:

1) The presence of electron-rich ligands in the coordination sphere of the metal increases the rate of oxidative-addition.

2. (cont.)

2) A low initial oxidation state of the metal is more favorable for oxidative—addition reactions to occur; e.g., all other factors being equal, Fe(0) is easier to oxidize than Co(I), which is easier to oxidize than Ni(II), even though the d—electron configuration is the same for these metals.

3) The tendency for oxidative—addition to occur increases down a given group, e.g. Ir(II) is easier to oxidize than Rh(II), which is easier to oxidize than Co(II).

4) Oxidative—addition occurs more readily in coordinatively unsaturated systems.

Steric effects have also been shown to be important. In general, valid predictions of relative reactivities can only be made between compounds of very similar structure.

a) $[Ir(dppe)_2]^+$

Collman, J.P.; Roper, W.R. Adv. Organomet. Chem. **1968**, 7, 53.

b) $RhCl(PPh_3)_3$

Collman, J.P.; Roper, W.R. Adv. Organomet. Chem. **1968**, 7, 53.

c) $IrCl(CO)(PPh_3)_2$

Collman, J.P.; Roper, W.R. Adv. Organomet. Chem. **1968**, 7, 53.

d) $Ir(CO)(PPh_3)_2Br$

Halpern, J. Accts. Chem. Res. **1970**, 3, 386.

e) $Rh(dmpe)_2^+$

Collman, J.P. Accts. Chem. Res. **1968**, 1, 136.

f) $Ir(PPh_3)_2(CO)Cl$

Halpern, J. Accts. Chem. Res. **1970**, 3, 386.

g) $Os(CO)_5$

Although addition of electron—donating groups enhances reactivity towards oxidative—addition, CO must dissociate first in this reaction. The presence of the phosphine ligands strengthens the Os—CO bond, thus preventing the dissociation necessary for the oxidative addition reaction to occur.

Collman, J.P.; Roper, W.R. Adv. Organomet. Chem. **1968**, 7, 53.

3.

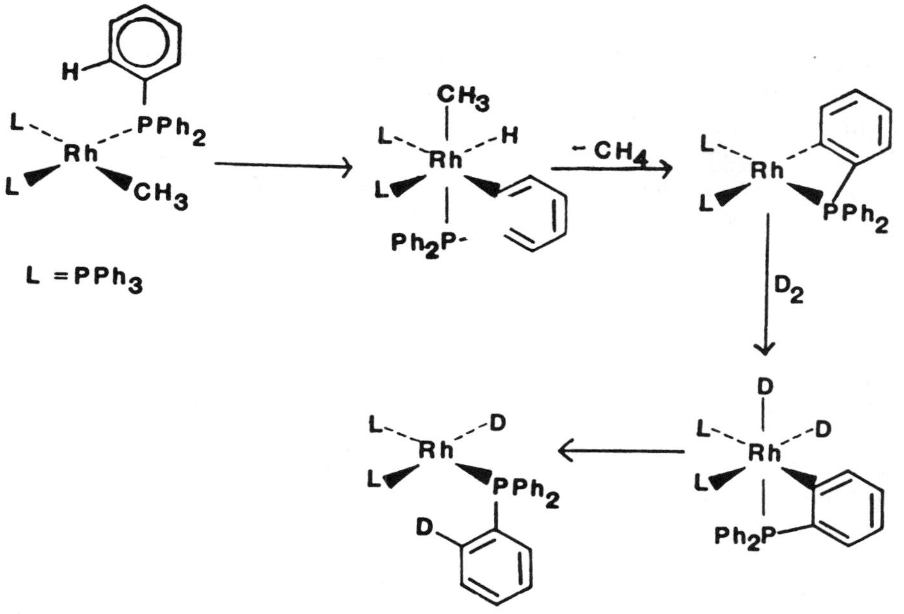

L = PPh$_3$

Keim, W. J. Organomet. Chem. 1969, 19, 161.

4. a) The independence of k_{obs} on [PMe$_3$] above a certain concentration suggests that there may be two possible reaction pathways.

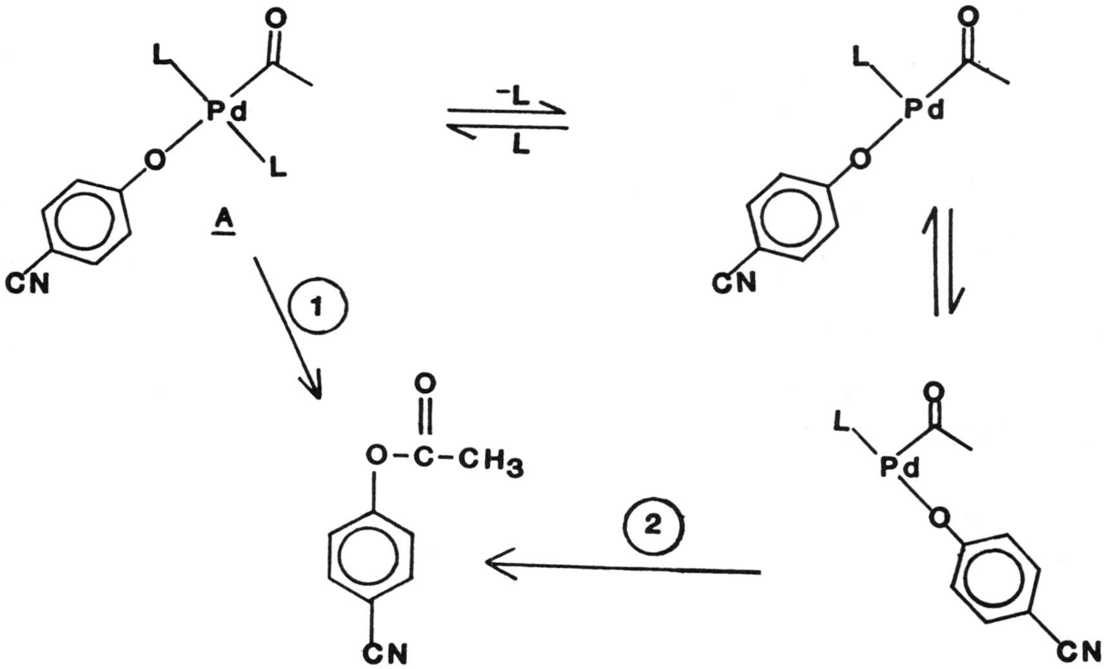

Pathway 2, involving initial dissociation of L and isomerization to a cis-acyloxy compound will be shut off with the addition of excess L, while pathway 1 will be operative at all concentrations of L. Direct reductive-elimination from the trans-acyloxy compound, A, (Pathway 1)

4. a) (cont.) seems unlikely, since reductive-elimination reactions do not occur readily from _trans_ compounds; however, it fits the observed kinetic data. Another possibility is isomerization of the four-coordinate _trans_ isomer to the _cis_ isomer followed by reductive-elimination. Kinetically, this is indistinguishable from the reductive-elimination from the four-coordinate _trans_ isomer.

b) The increase in k_{obs} when dmpe is added to the reaction mixture is thought to be due to displacement of both PEt_3 ligands by the chelating ligand, with a corresponding change in geometry to a _cis_-acyloxy complex, from which reductive-elimination is more facile.

Komiya, A.; Akai, Y.; Tanaka, K.; Yamamoto, T.; Yamamoto,A. Organomet. **1985**, _4_, 1130.

5.   **2** = _trans_-Pd(PPh$_3$)$_2$(Ph)Br, oxidative-addition

d) 6 = _cis_-Pd(PPh$_3$)$_2$(Ph)H

The last step is a reductive-elimination.

Tamaru, Y.; Yamada, Y.; Inoue, K.; Yamamoto, Y.; Yoshida, Z. _J. Org._ _Chem._ **1983**, _48_, 1286.

6. a) The mechanism involves a concerted addition of H_2 for reaction **1**.

Longato, B.; Morandini, F.; Bresadola, S. _Inorg._ _Chem._ **1976**, _15_, 650.

6. **a)** (cont.) For reaction 2, the mechanism involves an S_N2-type attack of the metal on methyltriflate.

b) Oxidative-addition to polar molecules with good leaving groups tend to favor S_N2-type reactions. H^- is a terrible leaving group compared to $[OSO_2CF_3]^-$.

A radical mechanism is also a possibility. To determine whether or not this is the case, one must:

1) check for inhibition by radical scavengers such as galvinoxyl or duroquinone and initiation by radical initiators such as benzoyl peroxide or 2, 2'-azobisisobutyronitrile (AIBN);

2) carry out the reaction using chiral substrates (alkyl halides or tosylates) and look for racemization of the products; and

3) determine the effect of the polarity of the solvent on the rate of the reaction. This effect should be minimal for a radical reaction.

Stang, P.J.; Schiavelli, M.D.; Chenault, H.K.; Breidegam, J.L. Organomet. 1984, 3, 1133 and references therein.

7. a) The mechanism of the reaction is shown below:

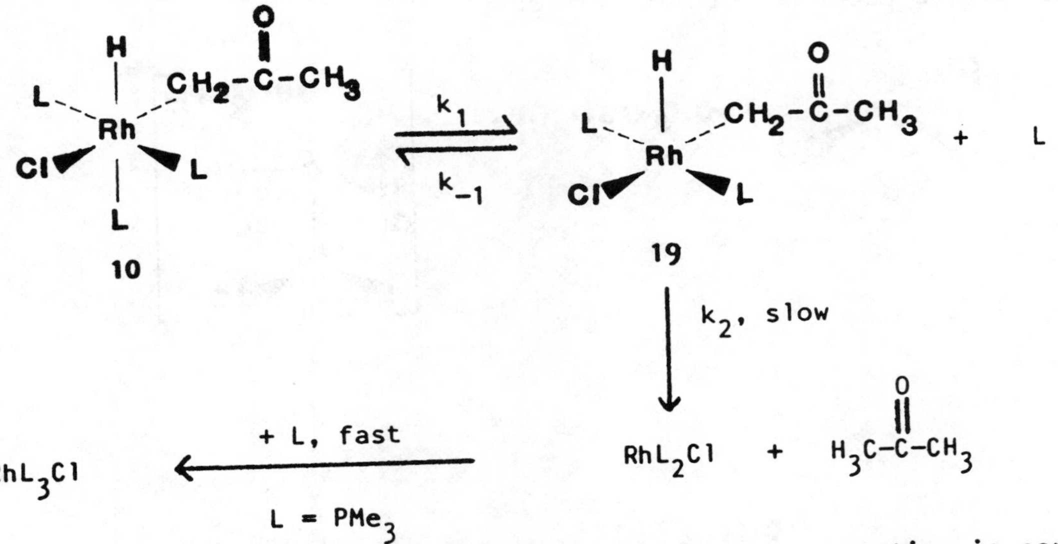

L = PMe$_3$

Because no crossover products are obtained when the reaction is carried out with 10-d$_0$ and 10-d$_6$, the reductive-elimination must be intramolecular. This conclusion is also supported by observation 2). The existence of an isotope effect indicates that reductive-elimination must occur at or before the rate-determining step. A dissociative first step is supported by observations 3) and 4), where addition of excess ligand <u>decreases</u> the rate and "phosphine sponge" <u>increases</u> the rate of the reaction. The linear variation in 1/k$_{obs}$ versus [PMe$_3$] is indicative of a rate law with [PMe$_3$] in the denominator.

b) The rate expression can be derived as follows:

d[acetone]/dt = k$_2$[19]

Using a steady state approximation for the intermediate **19**,

$$d[19]/dt = 0 = k_1[10] - k_{-1}[19][PMe_3] - k_2[19]$$

therefore, $k_1[10] = (k_{-1}[PMe_3] + k_2)[19]$

$$[19] = \frac{k_1[10]}{k_{-1}[PMe_3] + k_2}$$

$$d[acetone]/dt = \frac{k_2 k_1[10]}{k_{-1}[PMe_3] + k_2} = k_{obs}[10], \text{ at constant } [PMe_3]$$

Milstein, D. <u>Accts</u>. <u>Chem</u>. <u>Res</u>. 1984, <u>17</u>, 221.

8. **a)** Electrophilic cleavage reactions of this type are thought to occur via one or two electron oxidation steps involving formally Ru(III) or Ru(IV) intermediates. The ease with which these reactions occur is a function of the oxidation potential of the complex; thus, the lower the oxidation potential, the more facile the reaction. The compounds in the table can be placed in order of increasing reactivity based on their oxidation potentials: $CpRu(PPh_3)_2Me \sim CpRu(PPh_3)_2CH_2Ph >$ $CpRu(PPh_3)(CO)Me \sim CpRu(PPh_3)(CO)CH_2Ph > CpRu(CO)_2Me$. This trend in oxidation potentials is directly related to electron density at the metal. The addition of electron-rich phosphine ligands increases the electron density on the metal, thereby making the complex more susceptible to oxidation.

b) **1)** $CpRu(PPh_3)(CO)Cl + Ph-CH_3$

The proposed intermediate is

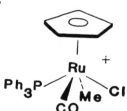

2) $CpRu(PPh_3)(CO)Cl + CH_3Cl$

The proposed intermediate is

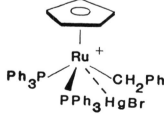

3) $CpRu(PPh_3)_2Br + Ph-CH_2-HgBr$

The proposed intermediate is

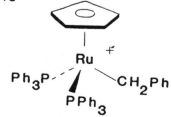

4) $CpRu(PPh_3)_2Br + 2 CuBr + Br-CH_2Ph$

The proposed intermediate is

8. b) (cont.) In the first three cases, electrophilic cleavage is proposed to occur via a two-electron oxidative-addition to give an 18-electron Ru(IV) intermediate. Reductive-elimination of R-X (R = CH_2Ph, CH_3; X = HgBr, Cl, H) to give a 16-electron $[CpRu(L)(L')]^+$ species, followed by coordination of Cl^- or Br^-, gives the observed products. In the oxidation involving copper halides, electrophilic cleavage is proposed to occur via an initial one-electron oxidation to give a 17-electron intermediate. The organic product $PhCH_2Br$ is then formed by nucleophilic attack by Br^- on the benzyl carbon atom. The resulting radical species $[CpRu(PPh_3)_2]^{\cdot}$ abstracts a halogen atom from $CuBr_2$ to give the observed organometallic product $CpRu(PPh_3)_2Br$.

Joseph, M.F.; Page, J.A.; Baird, M.C. *Organomet.* 1984, *3*, 1749.

9.

10.   **a)** Mechanism (2) is detailed below.

$$\underline{12} \quad \underset{k_{-1}}{\overset{k_1}{\rightleftharpoons}} \quad \underset{\underline{11}}{(dppe)Ir(CO)Br} \ + \ H_2 \quad \overset{k_2}{\longrightarrow} \quad \underline{13}$$

The rate expression is:

$$d[13]/dt = k_2[11][H_2]$$

Because $t_{1/2}$ for deuterium incorporation is fast relative to $t_{1/2}$ for isomerization, it is clear that **11** and **12** are in a rapid pre-equilibrium, and thus

$$K_{eq} = \frac{k_1}{k_{-1}} = \frac{[11][H_2]}{[12]}$$

Solving the equilibrium expression for [11] gives

$$[11] = \frac{k_1[12]}{k_{-1}[H_2]}$$

and substituting into the initial rate expression gives

$$\frac{d[13]}{dt} = k_2[H_2]\left(\frac{k_1[12]}{k_{-1}[H_2]}\right) = \frac{k_1 k_2[12]}{k_{-1}} = k_{obs}[12]$$

b) For the intramolecular rearrangement, the equation is:

$$12 \quad \overset{k}{\longrightarrow} \quad 13$$

and the rate is given by $d[13]/dt = k[12]$. This expression has the same form as that derived for the two step mechanism in part **a)**.

NOTE
The fact that deuterium incorporation is rapid relative to the isomerization process indicates that the coordinatively unsaturated intermediate **11** is a plausible intermediate in the mechanistic scheme; however, this is not sufficient evidence to eliminate an intramolecular pathway. There is literature precedent for both types of reactions (see reference). Thus, neither mechanism can be excluded by the data given.

c) The rate of deuterium incorporation could be monitored by either ^1H or ^2H NMR of the hydride region of the spectrum. The method actually used was ^1H NMR. Another possibility is to monitor the Ir-H(D) stretches in the infrared spectrum, although this may be difficult experimentally.

Johnson, C.E.; Eisenberg, R. J. Am. Chem. Soc. **1985**, _107_, 3148.

11. a)

$14 =$

$15 =$

Assignments for NMR data:
Complex **14**

Complex **15**

^{1}H NMR (δ, C_6D_6)

-16.88 (d, J_{P-H}=36 Hz, 1 H, Ir-H)
1.86 (dd, J_{P-H}=1.8, J_{H-H}=0.8 Hz,
   15 H, C_5Me_5)
1.24 (d, J_{P-H}=10.1 Hz, 9 H, PMe$_3$)
5.75, 6.95 (m, 2 H, -CH=C$\underline{H}_2$)
8.12 (ddd, J_{P-H}=3.3, $J_{H-H(cis)}$=10.2,
   $J_{H-H(trans)}$=17.5 Hz, -C$\underline{H}$=CH$_2$)

^{1}H NMR (δ, C_6D_6)

1.86 (d, J_{P-H}=1.5 Hz, 15 H,
   C_5Me_5)
1.01 (d, J_{P-H}=9.0 Hz, 9 H, PMe$_3$)
1.13, 1.27 (m, 4 H, CH$_2$=CH$_2$)

^{13}C (δ, C_6D_6, broad-band decoupled)

10.27 (s, $C_5\underline{Me}_5$)
18.82 (d, J_{P-C}=38.6 Hz, PMe$_3$)
92.50 (d, J_{P-C}=2.6 Hz, $\underline{C}_5Me_5$)
123.78 (d, J_{P-C}=3.0 Hz, -CH=$\underline{C}H_2$)
129.18 (d, J_{P-C}=13.4 Hz, -$\underline{C}H$=CH$_2$)

^{13}C (δ, C_6D_6, gated decoupled)

10.29 (q, J_{C-H}=126.1, $C_5\underline{Me}_5$)
10.55 (dd, J_{C-H1}=J_{C-H2}=150.6 Hz,
   CH$_2$=CH$_2$)
16.93 (dq, J_{P-H}=33.9, J_{C-H}=
   126.6 Hz, PMe$_3$)
90.19 (s, $\underline{C}_5Me_5$)

b) The mechanism of this reaction is not fully known; however, certain conclusions can be drawn about possible mechanisms from the knowledge of other systems. The first step is a reductive-elimination of cyclohexane to give the sixteen-electron intermediate Cp*Ir(PMe$_3$), **16**.

A logical second step would then be coordination of ethylene to **16** to give the ethylene complex **15**. oxidative-addition of a C-H bond of the bound ethylene could be envisioned as the next step, leading to the vinyl hydride **14**; however, the fact that **15** can not be independently converted to **14** makes this step unlikely. The authors propose the following mechanism involving two independent transition states, **17** and **18**, leading to the two different products **14** and **15**.

11. (cont.)

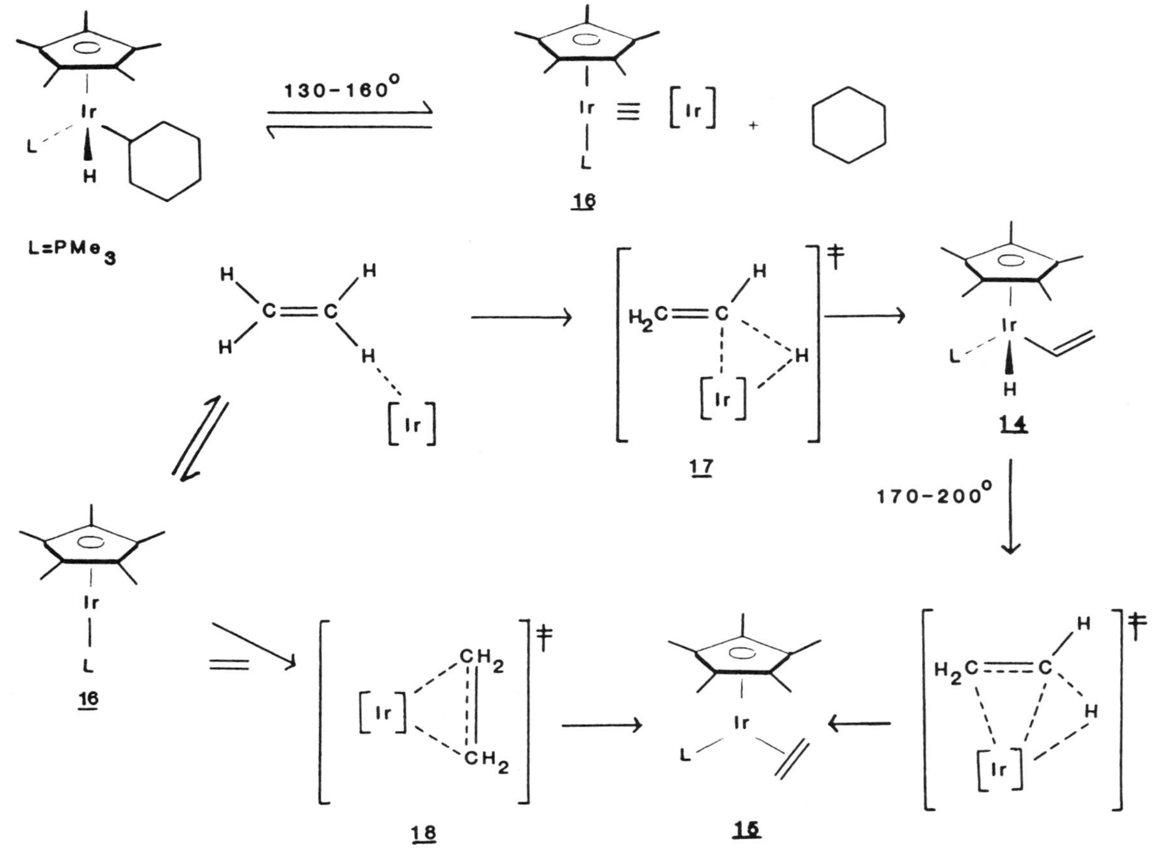

L=PMe₃

Stoutland, P.O.; Bergman, R.G. J. Am. Chem. Soc. 1985, 107, 4581.

6

Insertion Reactions

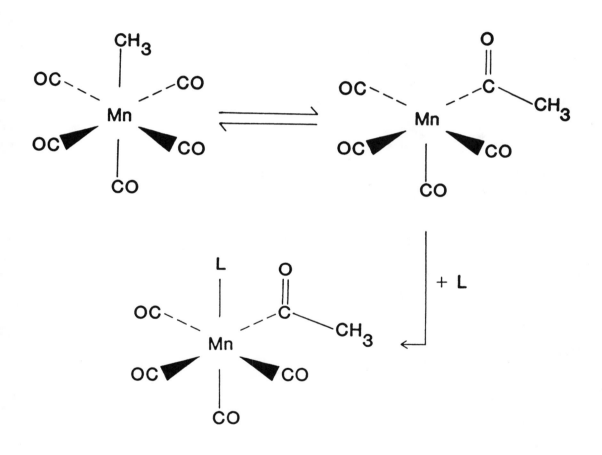

QUESTIONS

1. Predict the products of the following reactions.

a) $Cp(CO)_2Fe-CH_3$ + PPh_3 $\longrightarrow$

b) $Cp(CO)_3Mo-COCH_3$ $\xrightarrow{\Delta}$

c) $Cp_2(Cl)Ti-Et$ + CO $\longrightarrow$

d) $Et-Re(CO)_5$ + CH_3CN $\longrightarrow$

e) $cis-Et_2Os(CO)_4$ + 100 atm CO $\longrightarrow$

f) $(dppe)Pt(OMe)_2$ + 2 CO $\longrightarrow$

g) $(CO)_3(PPh_3)Co-CH_2Ph$ + ^{13}CO $\longrightarrow$

h) $[Cp_2Mo(H)(C_2H_4)]^+$ + PPh_3 $\longrightarrow$

2. For each group of insertion reactions given, predict which reaction will occur most rapidly. Explain your reasoning.

a) $Cp(CO)_3Cr-CH_3$ + PPh_3 $\longrightarrow$ $Cp(CO)_2(PPh_3)Cr-COCH_3$

$Cp(CO)_3Mo-CH_3$ + PPh_3 $\longrightarrow$ $Cp(CO)_2(PPh_3)Mo-COCH_3$

$Cp(CO)_3W-CH_3$ + PPh_3 $\longrightarrow$ $Cp(CO)_2(PPh_3)W-COCH_3$

b) $Cp(CO)_3Mo-CF_3$ + CO $\longrightarrow$ $Cp(CO)_3Mo-COCF_3$

$Cp(CO)_3Mo-CH_3$ + CO $\longrightarrow$ $Cp(CO)_3Mo-COCH_3$

c) $Cp(CO)_2Fe-CH_3$ + PPh_3 $\xrightarrow{THF}$ $Cp(CO)(PPh_3)Fe-COCH_3$

$Cp(CO)_2Fe-CH_3$ + PPh_3 $\xrightarrow{benzene}$ $Cp(CO)(PPh_3)Fe-COCH_3$

d) $[Fe(CO)_4Et]^-$ + PMe_3 $\longrightarrow$ $[Fe(CO)_3(PMe_3)COEt]^-$

$[Fe(CO)_4CH_2Ph]^-$ + PMe_3 $\longrightarrow$ $[Fe(CO)_3(PMe_3)COCH_2Ph]^-$

2. (cont.)

e) $(CO)_5Mn-CH_3$ + CO $\xrightarrow{AlBr_3}$ $(CO)_5Mn-COCH_3$

$(CO)_5Mn-CH_3$ + CO $\longrightarrow$ $(CO)_5Mn-COCH_3$

3. When **1** is treated with Ag^+, **2a** is formed. If **2a** is heated, a 50/50 mixture of **2a** and **2b** is generated. Propose a mechanism for the conversion of **2a** to **2b**.

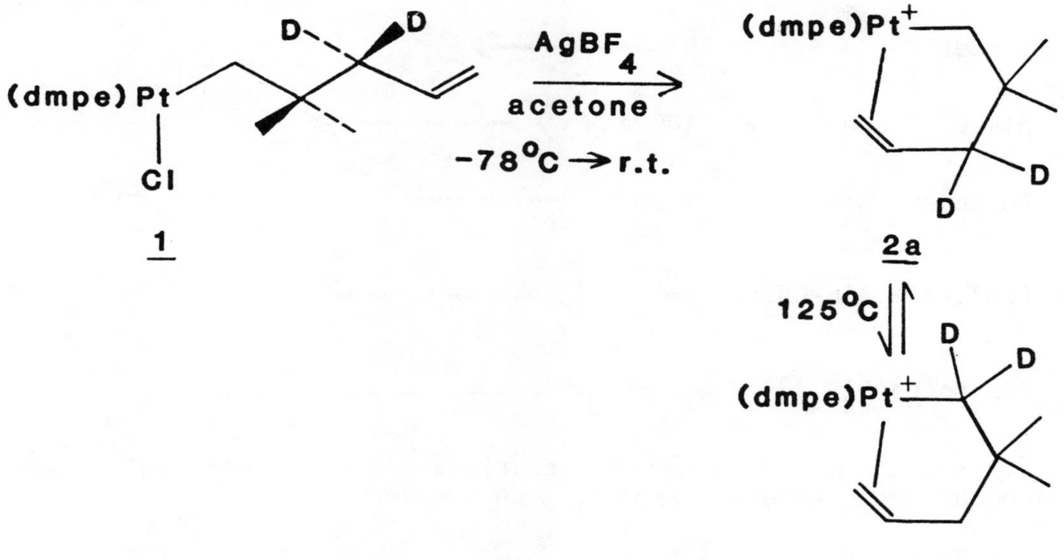

4. Upon standing at room temperature, compound **4** will decarbonylate to give **5**.

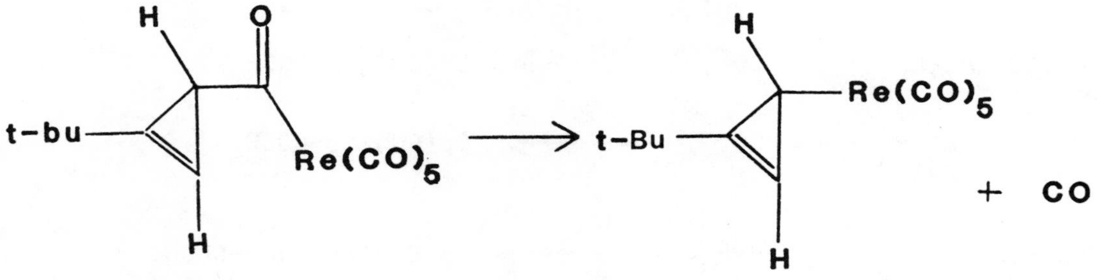

^1H NMR (δ, $CDCl_3$)

For **4**:
1.15 (s, 9 H)
2.92 (d, 1 H, J = 1.2 Hz)
6.18 (d, 1 H, J = 1.2 Hz)

For **5**:
1.20 (s, 9 H)
2.20 (d, 1 H, J = 1.2 Hz)
6.20 (d, 1 H, J = 1.2 Hz)

4. (cont.) Similarly, 4-d$_1$ will decarbonylate to 5-d$_1$. Based on the NMR spectral data shown below, give the position of the deuterium in 4-d$_1$ and 5-d$_1$.

4-d$_1$	5-d$_1$
1.15 (s, 9 H)	1.20 (s, 9 H)
2.92 (s, 1 H)	6.20 (s, 1 H)

5. Propose a mechanism for the following transformation.

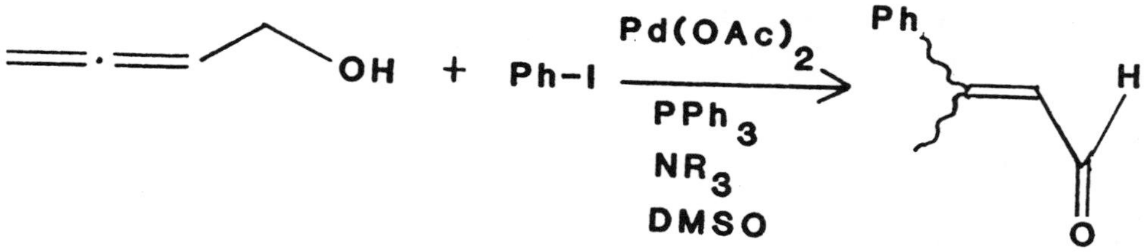

6. Predict the product and propose a mechanism for the following reaction.

HINT
Two insertion reactions occur.

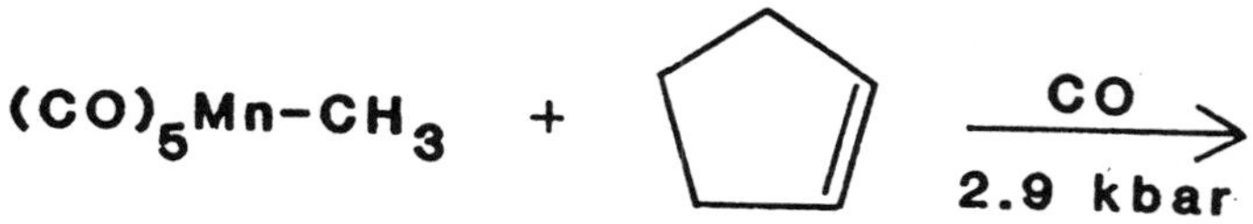

7. Treatment of CH$_3$Mn(CO)$_5$ with Ph$_2$PCH$_2$PPh$_2$ gives a product which has the formula C$_5$H$_3$MnO$_4$(Ph$_2$PCH$_2$PPh$_2$). The IR spectrum shows ν_{CO} at 1999, 1916 (broad), and 1588 cm^{-1}. The NMR spectrum shows, in addition to phenyl protons, a singlet at δ 2.34 and a twelve line pattern at ca. δ4.60. Deduce the structure and detailed stereochemistry of the product.

8. The following reaction has been observed

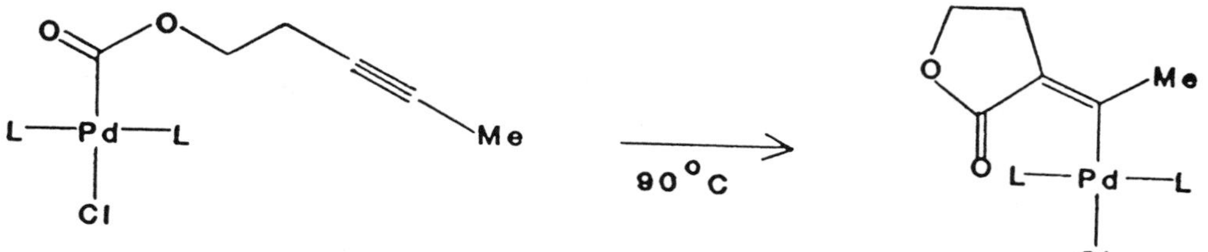

L = PPh$_3$

8. (cont.)

In addition, the following two observations have been made.

1) Addition of 0.5 eq of L decreases the rate by a factor of 50.

2) Without added phosphine, a plot of $-\ln (A-A_\infty)$ vs time shows upward curvature at short reaction times.

Propose a mechanism for this reaction. Be sure to show all intermediates. In addition, state which steps are fast, which are slow, and which are equilibria.

9. The following reactions are observed

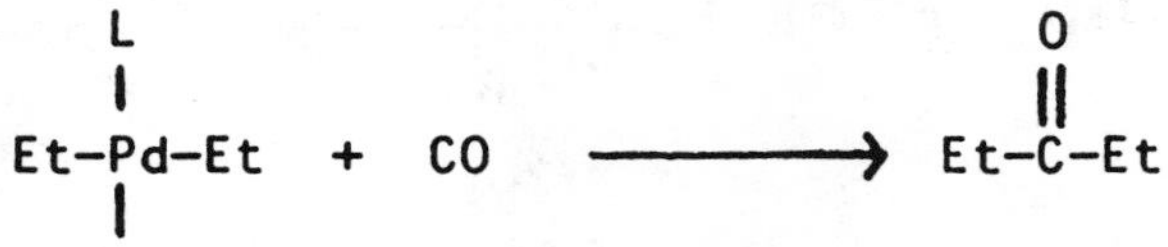

$$\underset{\textbf{14}}{\overset{\displaystyle L}{\underset{\displaystyle L}{\mid}} \text{Et-Pd-Et}} \quad + \quad CO \quad \longrightarrow \quad \underset{\textbf{15}}{\overset{\displaystyle O}{\text{Et-C-Et}}}$$

$$\underset{\textbf{16}}{\overset{\displaystyle L}{\underset{\displaystyle Et}{\mid}} \text{Et-Pd-L}} \quad + \quad CO \quad \longrightarrow \quad \underset{\textbf{17}}{\overset{\displaystyle O}{\text{Et-C-H}}} \quad + \quad \underset{\textbf{18}}{H_2C{=}CH_2}$$

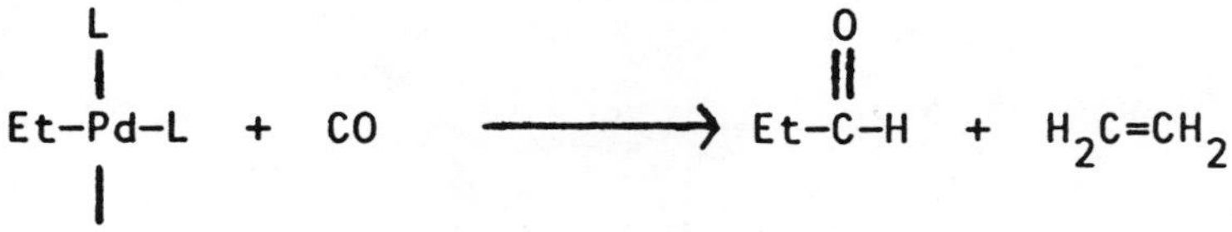

L = PMe$_2$Ph

Propose a mechanism(s) to account for the different behavior of the <u>cis</u> and <u>trans</u> isomers.

10. Supply plausible mechanisms for the following reactions:

a)

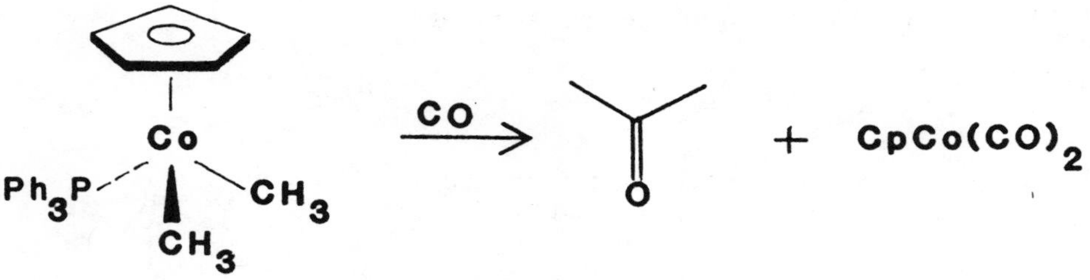

10. b)

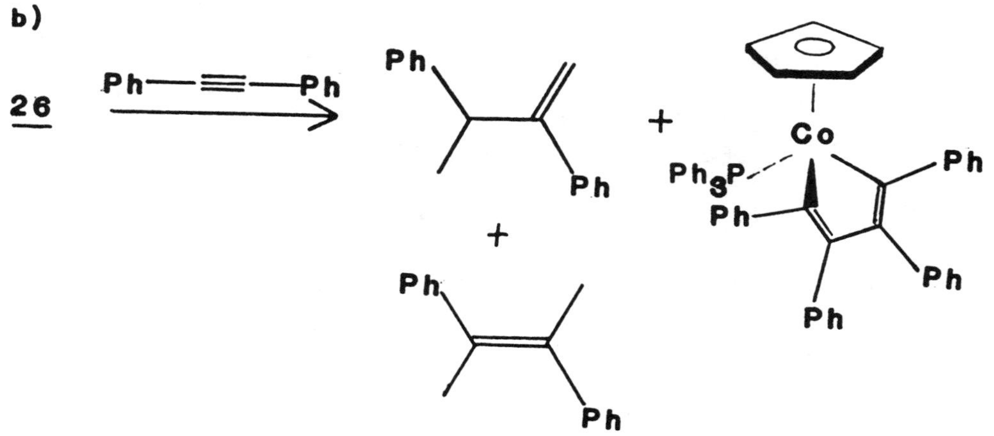

c)

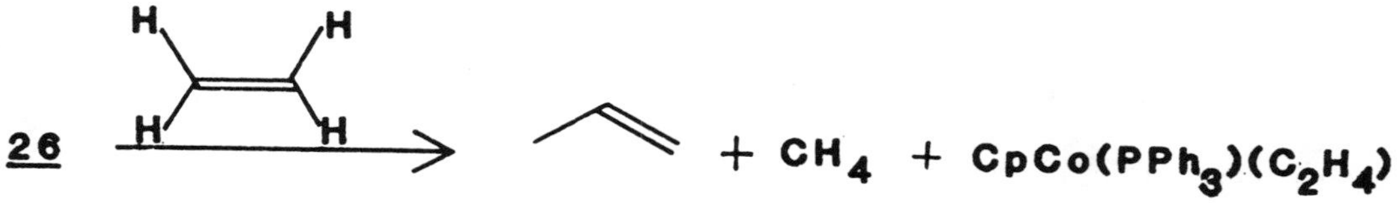

11. Propose a structure and assign all spectral data for product **27** arising from the following reaction.

$$H_3C-W(CO)_5^- \quad + \quad CO_2 \quad \longrightarrow \quad 27$$

Partial IR (ν, cm^{-1}, THF): 2059(w), 1955(w), 1907(s), 1842(m)

^{13}C NMR (δ, CD$_3$CN, broad-band decoupled): 205.0, 200.3, 175.7, 22.1

12. The following reaction occurs to give product **28**. On heating, **28** reacts further to give **29**.

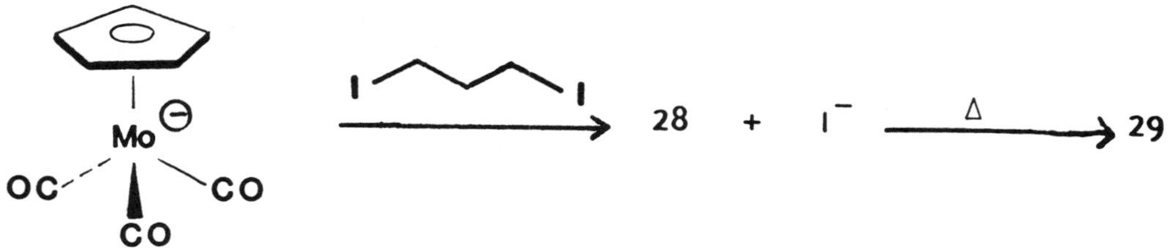

Spectral Data for **28**:

The ^{1}H NMR spectrum shows four different types of protons.

Spectral Data for **29**:

IR (ν_{CO}, CH$_2$Cl$_2$): 1985, 1909 cm^{-1}

12. (cont.)

^{1}H NMR (δ, CDCl$_3$):   5.57 (s, 5 H)
                                4.60 (t, J = 7 Hz, 2 H)
                                3.60 (t, J = 7 Hz, 2 H)
                                1.92 (quint., J = 7 Hz, 2 H)

^{13}C NMR (δ, CDCl$_3$, broad-band decoupled):   316.6
                                                         223.4
                                                          96.9
                                                          82.0
                                                          58.8
                                                          22.5

a) Identify **28** and **29** based on the spectral data given.

b) Postulate a mechanism for the conversion of **28** to **29**.

ANSWERS

1. a) Cp(CO)(PPh$_3$)Fe-COCH$_3$

Bibler, J.P.; Wojcicki, A. Inorg. Chem. **1966**, 5, 889.

b) Decarbonylation occurs to give Cp(CO)$_3$Mo-CH$_3$.

King, R.B.; Bisnette, M.B. J. Organomet. Chem. **1964**, 2, 15.

c) Carbon monoxide insertion occurs to give an η^2-acyl complex.

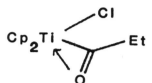

Fachinetti, G; Floriani, C. J. Organomet. Chem. **1974**, 71, C5.

d) (cis)-(CH$_3$CN)Re(CO)$_4$(COEt)

Martin, B.D.; Warner, K.E.; Norton, J.R. J. Am. Chem. Soc. **1986**, 108, 33.

e) cis-(EtCO)$_2$Os(CO)$_4$

L'Eplattenier, F.; Pelichet, C. Helv. Chim. Acta **1970**, 53, 1091.

f) (dppe)Pt(CO$_2$Me)$_2$

Bryndza, H.E. Organomet. **1985**, 4, 1686.

g) (^{13}CO)(CO)$_2$(PPh$_3$)Co-COCH$_2$Ph

Nagy-Magos, Z.; Bor, G.; Marko, L. J. Organomet. Chem. **1968**, 14, 205.

h) [Cp$_2$Mo(Et)(PPh$_3$)]$^+$

Benfield, F.W.S.; Green, M.L.H. J. Chem. Soc., Dalt. Trans. **1973**, 1848.

2. a) Because the "insertion" reaction actually involves a methyl migration and a concomitant breaking of the metal-methyl bond, the strength of that bond will affect the rate of the reaction. The relative strengths of the metal-carbon bonds are W-C >> Mo-C > Cr-C, thus the insertion reaction will proceed most rapidly for the chromium complex.

b) Electron-withdrawing substituents on the migrating alkyl group decrease the rate of migration; thus, the trifluoromethyl complex will undergo insertion at a slower rate.

c) Insertion reactions that generate coordinatively unsaturated intermediates generally proceed faster in a coordinating solvent, since the intermediate can be stabilized by interaction with the solvent; thus, the reaction carried out in THF should proceed at a faster rate.

2. **d)** Electron-withdrawing substituents on the migrating alkyl group decrease the rate of migration; thus, the benzyl complex will undergo insertion at a slower rate.

e) The rate of migratory insertion is increased in the presence of Lewis acids which stabilize the acyl intermediate via coordination to the acyl oxygen; thus, the reaction carried out in the presence of $AlBr_3$ will proceed faster.

3. The conversion of **2a** to **3** is an intramolecular olefin insertion into the metal-alkyl bond.

In intermediate **3**, the 1,4-bond can break to give back **2a** or the 3,4-bond can break to give **2b**. In the absence of the deuterium label, these two bonds are related by symmetry so there is a 50/50 chance of either bond breaking. This analysis assumes that there is no secondary isotope effect on the bond-breaking reaction when D is substituted for H.

Flood, T.C.; Bitler, S.P. J. Am. Chem. Soc. **1984**, _106_, 6076.

4. For this compound, the decarbonylation occurs with an allylic rearrangement. This reaction represents the first known case of this type of rearrangement.

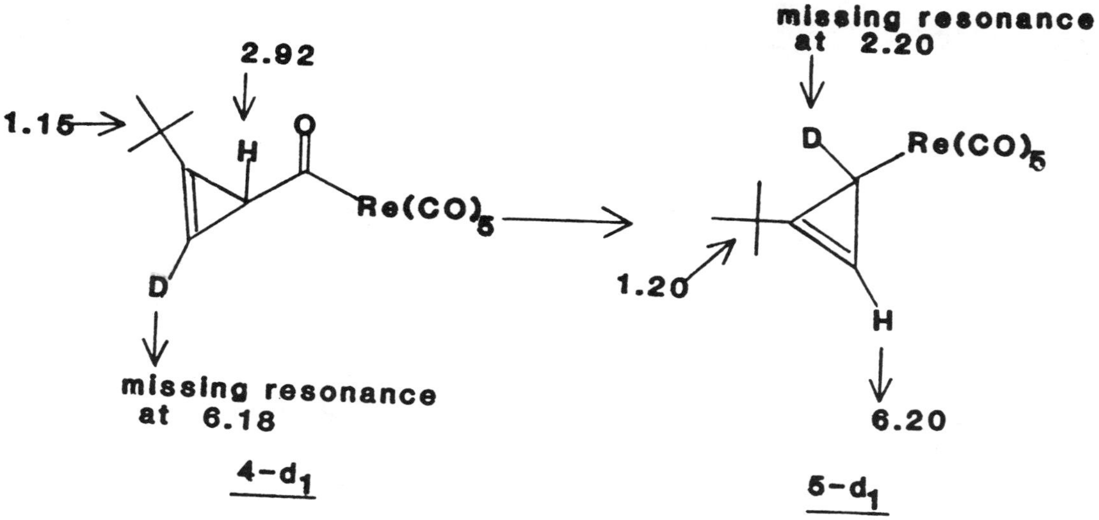

DeSimone, D.M.; Desrosiers, P.J.; Hughes, R.P. J. Am. Chem. Soc. 1982, 104, 4842.

5. In the mechanistic scheme shown below, "Pd" represents the metal and an unknown number of ancillary ligands.

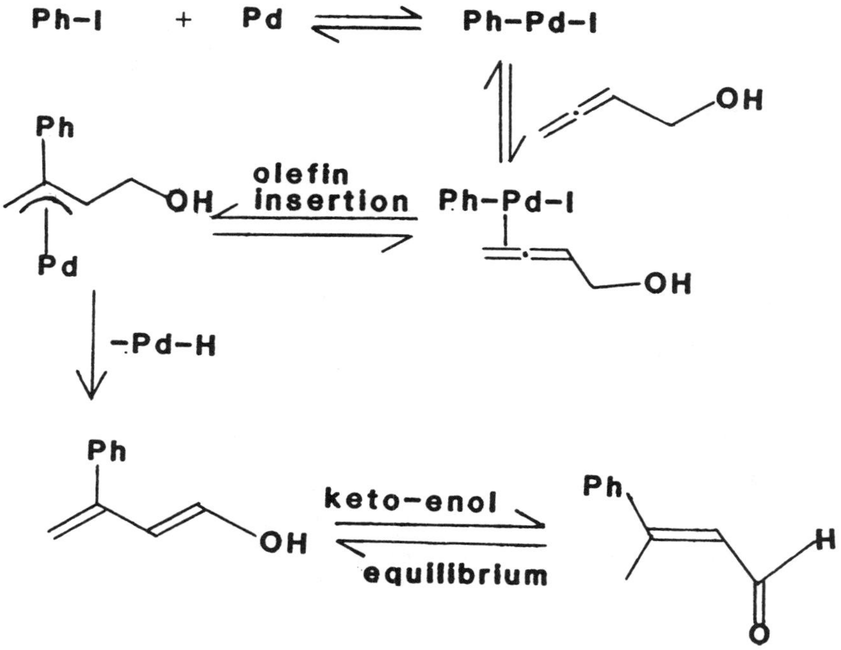

Shimizu, I.; Sugiura, T.; Tsuji, J. J. Org. Chem. 1985, 50, 537.

6. This reaction involves two insertion reactions. The first is a CO insertion and the second an olefin insertion.

6. (cont.)

$$H_3C-Mn(CO)_5 \underset{}{\overset{CO}{\rightleftharpoons}} H_3C-\overset{\overset{\displaystyle O}{\|}}{C}-Mn(CO)_5 \rightleftharpoons H_3C-\overset{\overset{\displaystyle O}{\|}}{C}-Mn(CO)_4$$

The product shows no tendency toward further CO or olefin insertions.

DeShong, P.; Slough, G.A. Organomet. **1984**, **3**, 636.

7. The IR spectrum shows terminal carbonyls and an acyl, implying that a CO insertion has taken place. Thus, the product is the six-coordinate octahedral complex

$$H_3C-\overset{\overset{\displaystyle O}{\|}}{C}-Mn(CO)_3(Ph_2PCH_2PPh_2)$$

The two phosphorus atoms of the chelating phosphine must be cis, so the question is whether the acyl ligand is trans to a CO (**6**) or trans to a phosphine (**7**).

6   **7**

Each molecule has only one mirror plane. In **6**, H_1 and H_2 are equivalent because the mirror plane passes between them. On the other hand, the phosphorous atoms are not equivalent, so each should split the proton signal into a doublet thus giving a doublet of doublets or 4 peaks total.

In **7**, H_1 and H_2 are not equivalent since they lie _in_ the mirror plane. Each will split the other into a doublet. Here the two phosphorus atoms are equivalent and will split each proton signal into a triplet (this is for the same reason that in e.g. CH_2CH_3, the methyl is split into a triplet by the two equivalent hydrogens on the methylene). For

7. (cont.) the <u>trans</u> isomer, the total number of lines expected for H_1 should include a triplet due to splitting by the equivalent phosphorous atoms, and a doublet due to H_2. This will result in a doublet of triplets or a six-line pattern. Secondly, a six-line pattern is observed for H_2, or a total of twelve peaks. This is consistent with the observed spectrum, so the acyl is <u>trans</u> to a CO.

Kraihanzel, C.S.; Maples, P.K. <u>J</u>. <u>Organomet</u>. <u>Chem</u>. 1969, <u>20</u>, 269.

8. To form the product, the acetylene must insert into the metal–acyl bond. This could occur from either a four-coordinate intermediate such as **10** or a five-coordinate intermediate such as **12**. Inhibition by phosphine does not distinguish between these two possibilities, since added L may stop the acetylene from coordinating to the metal in both cases.

Four-Coordinate Intermediate

Five-Coordinate Intermediate

With the addition of 0.5 eq of L, the equilibrium between **8** and **11** can only decrease the initial rate by a factor of two, and this is only true if **11** is highly favored in the equilibrium. Thus, the four-coordinate intermediate fits the data better. The proposed mechanism is shown on the following page.

8. (cont.)

We must now ask if **10** is a steady state intermediate. In other words, is $k_1 \ll k_{-1} + k_2$ or is $k_2 \ll k_1$ and k_{-1}?

The upward curvature in the plot of $-\ln(A-A_\infty)$ vs time indicates that the initial rate is slower than the final rate. We know that phosphine will decrease the rate of the reaction. If **10** were a steady-state intermediate, then the amount of free L generated in the reaction is negligible. On the other hand, if k_1/k_{-1} were a rapid equilibrium with k_2 as the slow step, a significant quantity of free L will be generated to slow down the reaction. As the reaction proceeds, the amount of free L will decline as it becomes coordinated to the product and the rate of reaction will increase. Thus, there must be a fast equilibrium followed by a rate determining insertion reaction.

Samsel, E.G.; Norton, J.R. J. Am. Chem. Soc. 1984, 106, 5505.

9.   The authors propose that all reactions have configurationally stable three-coordinate intermediates. However, they say that other possibilities involving five-coordinate intermediates cannot be excluded.

9. (cont.)

$$
\begin{array}{c}
\text{L} \\
| \\
\text{Et-Pd-Et} \\
| \\
\text{L}
\end{array}
\qquad
\overset{-\text{L}}{\underset{}{\rightleftharpoons}}
\qquad
\begin{array}{c}
\text{L} \\
| \\
\text{Et-Pd-Et}
\end{array}
$$

14   **19**

$$\Big\updownarrow + CO$$

$$
\begin{array}{c}
\text{L} \\
| \\
\text{Et-Pd} \\
| \\
\text{COEt}
\end{array}
\qquad
\overset{\text{alkyl}}{\underset{\text{migration}}{\rightleftharpoons}}
\qquad
\begin{array}{c}
\text{L} \\
| \\
\text{Et-Pd-Et} \\
| \\
\text{CO}
\end{array}
$$

21   **20**

$$
\begin{array}{c}
\text{O} \\
\|\ \\
\text{Et-C-Et}
\end{array}
$$

15

The _cis_ alkyl and acyl on **21** can reductively eliminate to give the ketone, **15**.

On the other hand, the _trans_ alkyl and acyl on **24** are too far apart to reductively eliminate. Since there is an open coordination site next to the alkyl, the metal is able to abstract the β-hydrogen to give a complex (**25**) which can easily form ethylene and the aldehyde.

$$
\begin{array}{c}
\text{L} \\
| \\
\text{Et-Pd-L} \\
| \\
\text{Et}
\end{array}
\ \overset{-\text{L}}{\rightleftharpoons}\
\begin{array}{c}
\text{L} \\
| \\
\text{Et-Pd} \\
| \\
\text{Et}
\end{array}
\ \overset{\text{CO}}{\rightleftharpoons}\
\begin{array}{c}
\text{L} \\
| \\
\text{Et-Pd-CO} \\
| \\
\text{Et}
\end{array}
$$

16   **22**   **23**

$$\Big\updownarrow \substack{\text{alkyl} \\ \text{migration}}$$

$$
\begin{array}{c}
\|\quad\ \text{O} \\
\text{—Pd—C} \\
|\quad\quad \text{Et} \\
\text{H} \quad \mathbf{25}
\end{array}
\ \overset{\substack{\text{β-hydrogen} \\ \text{elimination}}}{\longleftarrow}\
\begin{array}{c}
\quad\quad\quad\ \text{L} \quad \text{O} \\
\quad\quad\quad\ |\quad \| \\
\text{CH}_2\text{—CH}_2\text{—Pd—C} \\
\quad\quad\quad\quad\quad\quad \text{Et} \\
\text{H} \quad\quad\quad\quad \mathbf{24}
\end{array}
$$

$$
\begin{array}{c}
\text{O} \\
\|\ \\
\text{Et—C—H}
\end{array}
\ +\
\ \|\
$$

17   **18**

9. (cont.) Ozawa, F.; Yamamoto, A. <u>Chem</u>. <u>Lett</u>. 1981, 289.

For further information regarding three—coordinate group 10 metal complexes, see: Thorn. D.L.; Hoffmann, R. <u>J</u>. <u>Am</u>. <u>Chem</u>. <u>Soc</u>. 1978, <u>100</u>, 2079.

10. a)

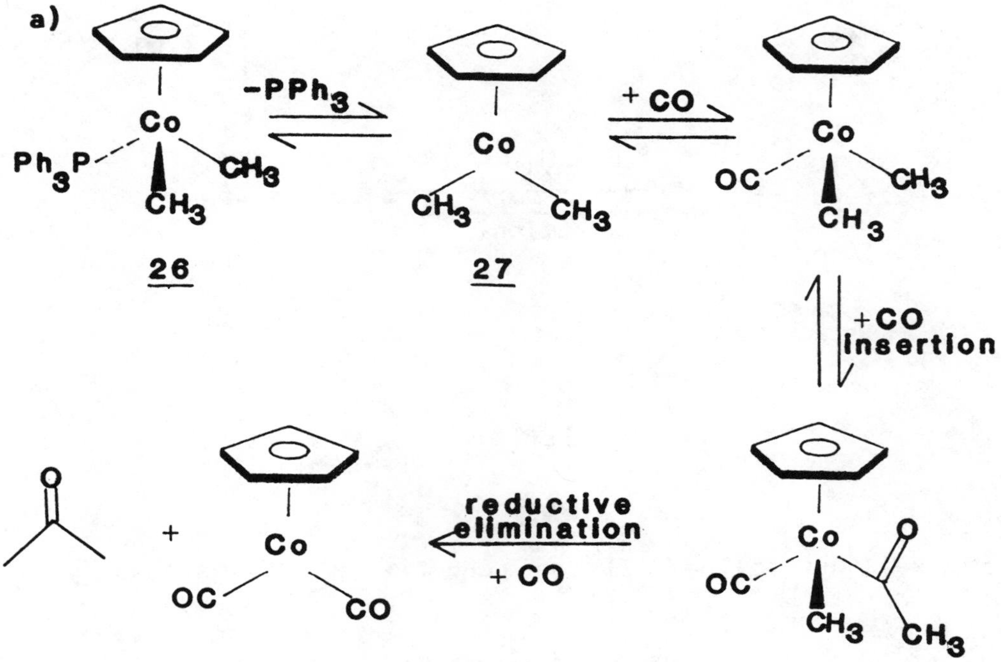

b)

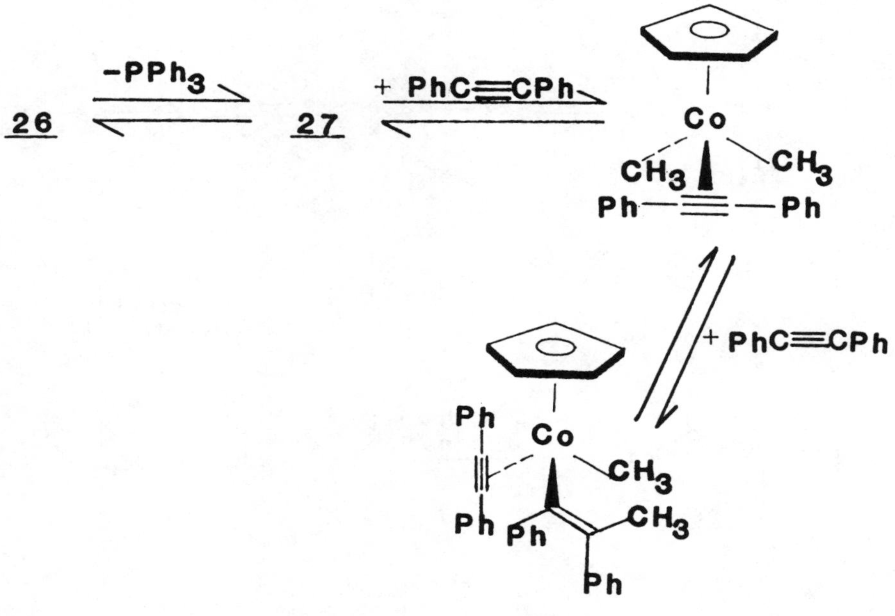

10. b) (cont.)

10. c)

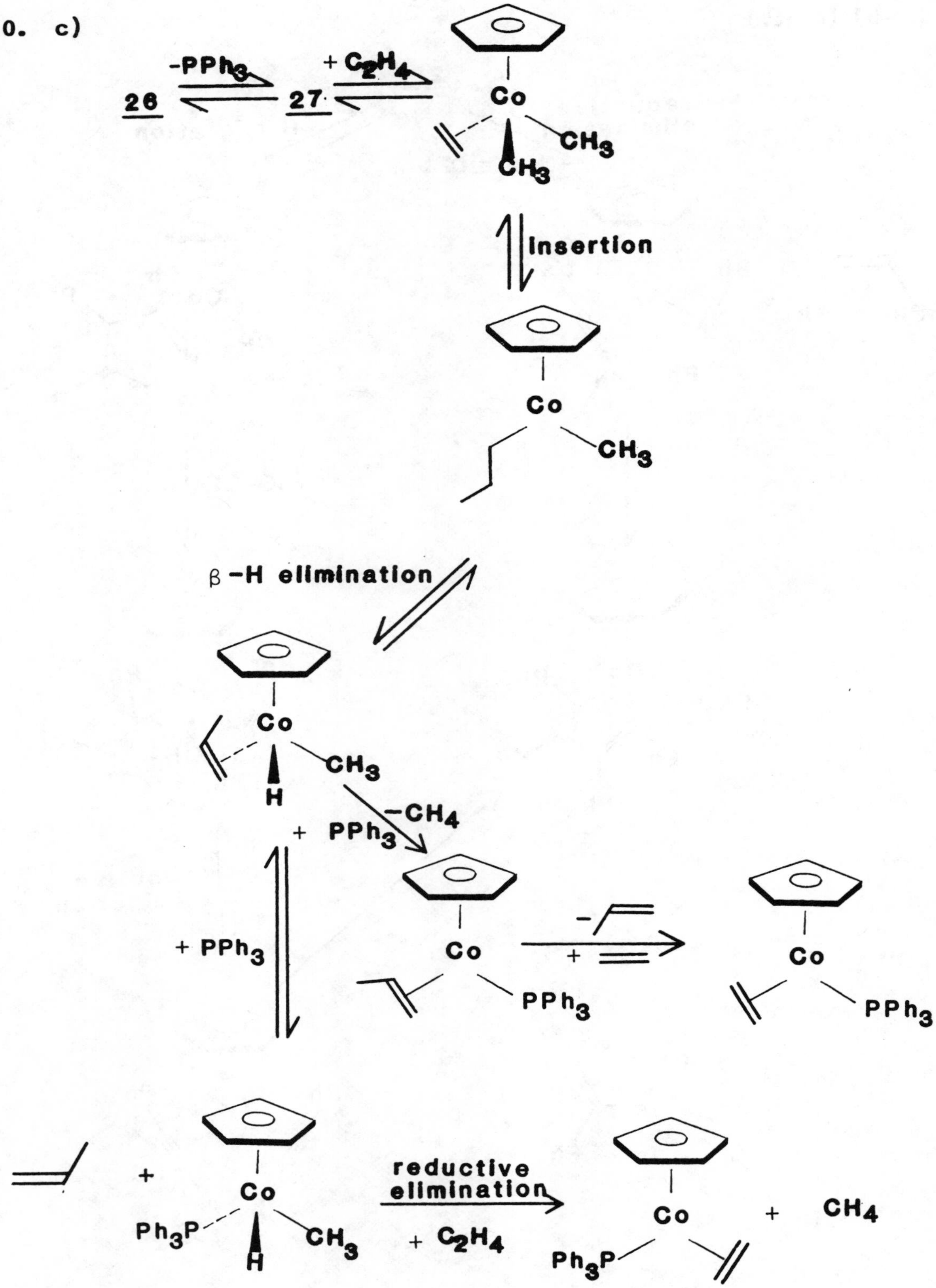
-PPh3
+C2H4
26
27
Co
CH3
CH3
insertion
Co
CH3
β-H elimination
Co
CH3
H
-CH4
+ PPh3
+ PPh3
Co
PPh3
+
Co
PPh3
+
Co
Ph3P
CH3
H
reductive
elimination
+ C2H4
Co
Ph3P
+
CH4

Bergman, R.G. Accts. Chem. Res. 1980, 13, 113.

11. The product is the result of CO_2 insertion into a tungsten—methyl bond.

$$[H_3C-\overset{\overset{\displaystyle O}{\|}}{C}-O-W(CO)_5]^-$$

The ^{13}C NMR data are assigned as follows:

δ 22.1 $-CH_3$

δ 175.7 $-\overset{\overset{\displaystyle O}{\|}}{C}-O-W$

δ 205.0   axial CO

δ 200.3   equatorial CO

The δ 175.7 resonance is a typical value for the carbonyl of a carboxylic acid. The peaks in the infrared spectrum are assigned to the terminal carbonyls in the anionic product.

Darensbourg, D.J.; Kudaroski, R.A. Adv. Organomet. Chem. **1983**, <u>22</u>, 129.

12. a)

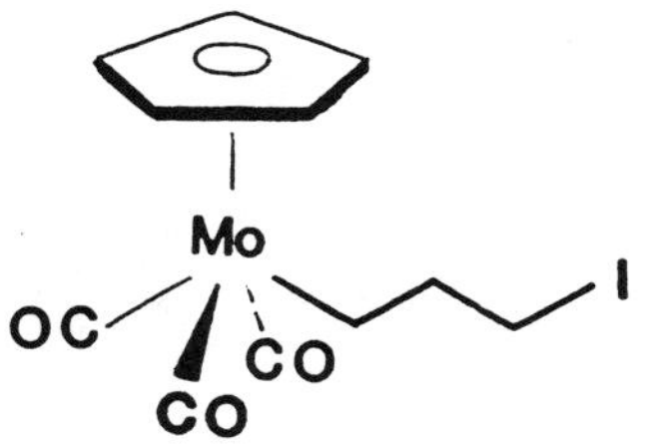

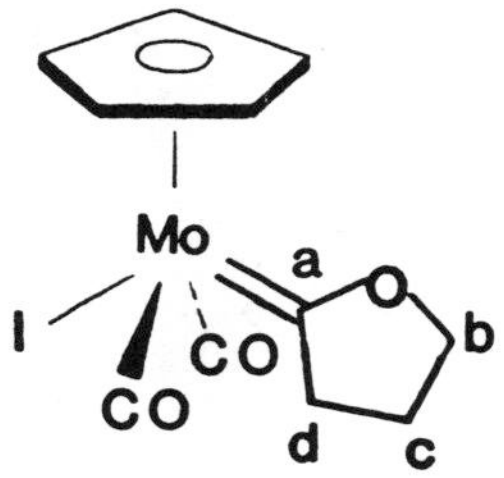

Assignments for NMR data for **29**:

^1H NMR

5.57 (s, 5 H, Cp)

4.60 (t, J = 7 Hz, 2 H, H_b)

3.60 (t, J = 7 Hz, 2 H, H_d)

1.92 (quint., J = 7 Hz, 2 H, H_c)

^{13}C NMR

316.6 (C_a)

223.4 (CO carbons)

96.9 (Cp carbons)

82.0 (C_b)

58.8 (C_d)

22.5 (C_c)

12. b)

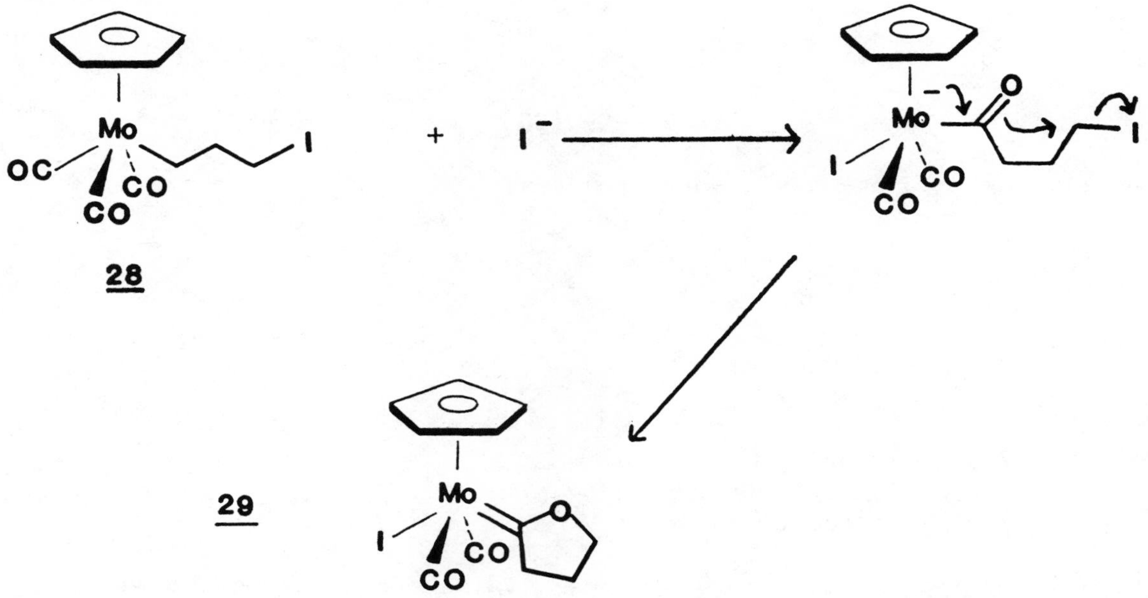

28

29

Bailey, N.A.; Chell, P.L.; Mukhopadhyay, A.; Tabbron, H.E.; Winter, M.J. J. Chem. Soc., Chem. Comm. 1982, 215.

7

Nucleophilic Attack on Coordinated Ligands

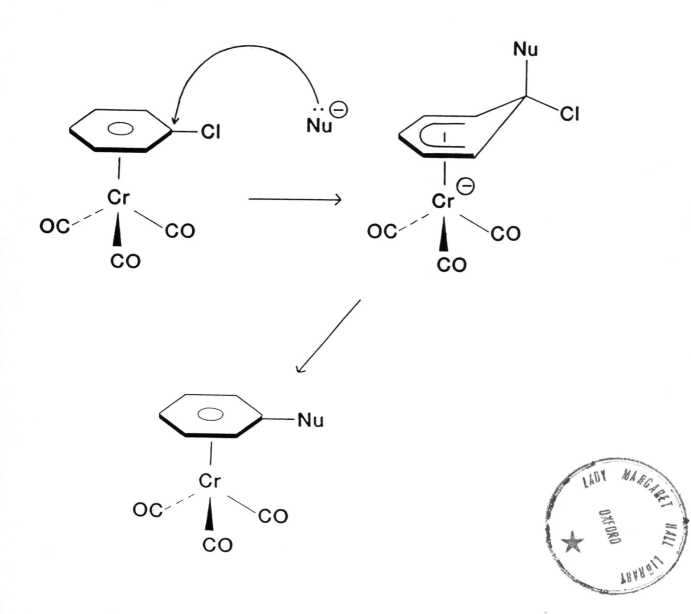

QUESTIONS

1. Predict the products of the following reactions.

a) $[CpRe(CO)_2(NO)]^+$ + $K(i-PrO)_3BH$ $\xrightarrow{\hspace{3cm}}$ 1

^1H NMR of 1 (δ, THF-d_8): 16.29 (s, 1 H), 5.87 (s, 5 H)

IR (THF): 1991, 1723, 1619 cm^{-1}

b) $[CpRe(CO)_2(NO)]^+$ + NaOH $\xrightarrow{\hspace{3cm}}$ 2

^1H NMR of 2 (δ, CD$_3$CN): 8.6 (br, 1 H), 5.84 (s, 5 H)

IR (Nujol): 2960, 2860, 2705, 2690, 1985, 1715, 1585 cm^{-1}

c) $(CO)_5Cr=C(OMe)(Me)$ + PR_3 $\xrightarrow{\hspace{3cm}}$ 3

d) $Cl_2Pd(AsPh_3)(C\equiv N-C_6H_4-p-CH_3)$ + $p-CH_3-C_6H_4-NH_2$ $\xrightarrow{\hspace{2cm}}$ 4

IR of 4: ν_{CN} 1536 cm^{-1}

e) $[Cp(CO)(P(OPh)_3)Fe(Me-C\equiv C-Me)]^+$ + NaSPh $\xrightarrow{\hspace{2cm}}$ 5

^1H NMR of 5 (δ, CDCl$_3$): 7.20 (m, 20 H), 4.37 (s, 5 H), 2.5 (br s, 3 H), 2.28 (s, 3 H)

f) $(\eta^6-C_6H_6)Cr(CO)_3$ + $LiCH_2CN$ $\xrightarrow{\hspace{3cm}}$ 6

2. Using the rules of Davies, Green, and Mingos (Tetrahedron **1978**, **34**, 3047), predict the kinetically-controlled product of nucleophilic attack on the unsaturated ligands of the following complexes.

a)

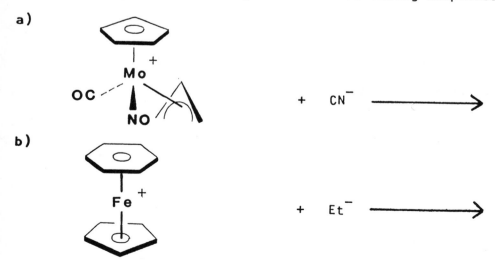

+ CN^- $\xrightarrow{\hspace{2cm}}$

b)

+ Et^- $\xrightarrow{\hspace{2cm}}$

176

2. c)

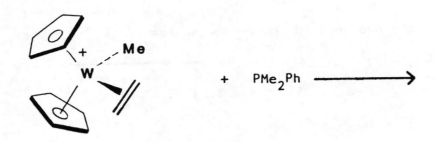

d)

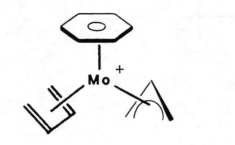   

e)

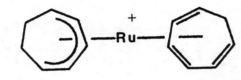   

f)

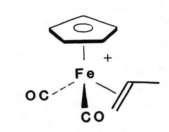   

3. The following reaction is observed.

$$Fe(CO)_5 \;+\; Me_3NO \xrightarrow[-30^\circ C]{THF} 7 \;+\; CO_2$$

Spectral data for **7**:

M/e (M$^+$) = 227

IR (cm^{-1}): 2050, 1960, 1940–1920

^{1}H NMR (δ, C$_6$D$_6$): 1.87 (s)

3. (cont.)

^{13}C NMR (δ, C_6D_6):   217.5 (s)
                                61.4 (q, J = 137 Hz)

Identify **7** and propose a mechanism for its formation.

4. Propose a mechanism for the following catalytic conversion.

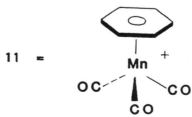

5. When heated in the presence of an added ligand, L (usually a phosphine), complex **8** reacts to give compound **9** by CO-insertion or compound **10** by insertion of the aryl into the manganese-methyl bond.

a) Propose a mechanistic scheme which accounts for the formation of these two products.

b) Derive a kinetic rate expression for formation of each of these products.

c) Experimentally, it is found that at high [L], **9** is formed exclusively, and at low [L], **10** is formed exclusively. How do you have to change your kinetic expression to fit these experimental results? In other words, what do these results imply about the relative rates of reactions?

6. Based on the spectral data, determine the product of each of the following reactions.

11 =

6. (cont.)

a)   **11** $\xrightarrow[\text{2) }H_2O\text{ work-up}]{\text{1) MeLi, THF, }-30^\circ C}$ **12**

IR(cm^{-1}): 2023, 1950, 1940

^{1}H NMR (δ, CDCl$_3$):

    0.4 (d, 3 H)
    2.5 (m, 1 H)
    3.1 (t, 2 H)
    4.6 (t, 2 H)
    5.6 (tt, 1 H)

b)   **11** $\xrightarrow[\text{2) }H_2O\text{ work-up}]{\text{1) LiMe}_2\text{Cu, }-78^\circ C\text{, ether}}$ **13**

IR(cm^{-1}): 1965, 1915, 1630

^{1}H NMR (δ, C$_6$D$_6$):

    2.60 (br s, 3 H)
    4.52 (s, 6 H)

c)   **11** $\xrightarrow[\text{2) }H_2O\text{ work-up}]{\text{1) LiMe}_2\text{Cu, }0^\circ C\text{, ether}}$ **14**

IR(cm^{-1}): 1975, 1930

^{1}H NMR (δ, C$_6$D$_6$):

    0.17 (s, 3 H)
    4.40 (s, 6 H)

7. The following reaction sequence can be used to make β-lactams.

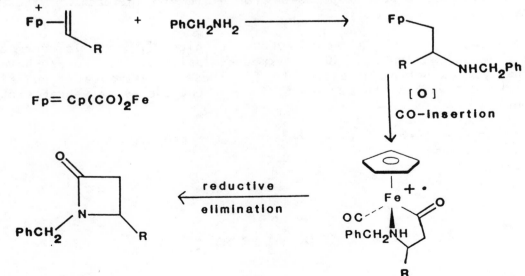

Fp = Cp(CO)$_2$Fe

7. a) Assuming <u>cis</u> (internal) nucleophilic attack and retention of configuration in the CO-insertion step, predict the stereochemistry of the product of the reaction of $[Cp(CO)_2Fe(\underline{cis}\text{-}2\text{-butene})]+$ and benzylamine.

b) Now assume <u>trans</u> (external) nucleophilic attack and predict the stereochemistry of the resulting β-lactam.

c) Using the NMR data given below for **15a** and **15b**, determine which mode of attack, external (<u>trans</u>) or internal (<u>cis</u>), prevails for the reaction:

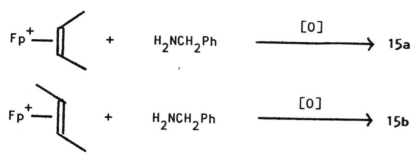

^{1}H NMR data (δ, CDCl$_3$)

For **15a:**

7.28 (s, 5 H)
4.62 (d, J = 15 Hz, 1 H)
4.05 (d, J = 15 Hz, 1 H)
3.16 (dq, J = 6, 2 Hz, 1 H)
2.74 (dq, J = 6, 2 Hz, 1 H)
1.25 (d, J = 6 Hz, 3 H)
1.17 (d, J = 6 Hz, 3 H)

For **15b:**

7.23 (m, 5 H)
4.52 (d, J = 15.5 Hz, 1 H)
4.00 (d, J = 15.5 Hz, 1 H)
3.51 (m, J = 6.3, 6.0 Hz, 1 H)
3.13 (m, J = 7.5, 6.0 Hz, 1 H)
1.11 (d, J = 7.5 Hz, 3 H)
1.01 (d, J = 6.3 Hz, 3 H)

8. For the sequence of reactions shown below, propose structures for compounds **16**, **17**, and **18**. Explicitly indicate all regiochemistry and stereochemistry.

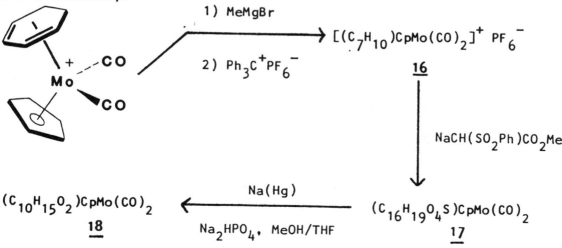

8. (cont.) Spectral Data for **18**:

IR (CCl$_4$): 1945, 1870, 1740 cm^{-1} (major peaks)

^{1}H NMR (δ, CDCl$_3$):   0.65 (d, J = 14.5 Hz, 1 H)
0.85 (dm, J = 14.5 Hz, 1 H)
1.22 (d, J = 7.2 Hz, 3 H)
1.90 (m, 1 H)
2.30 (m, 1 H)
2.55 (d, J = 6.8 Hz, 2 H)
3.5–3.75 (m, 2 H)
3.69 (s, 3 H)
4.17 (t, J = 7.0 Hz, 1 H)
5.29 (s, 5 H)

Assign all peaks in the IR and NMR spectra.

9. Give structures for compounds **19**, **20**, **21** and rationalize all spectral data.

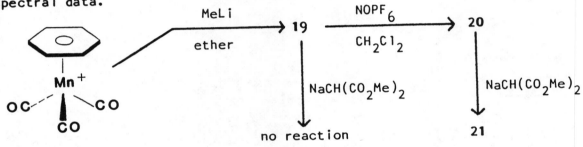

19: ^{1}H NMR (δ, CDCl$_3$):   0.4 (d, 3 H)
2.5 (m, 1 H)
3.1 (t, 2 H)
4.6 (t, 2 H)
5.6 (tt, 1 H)

partial IR (cm^{-1}): 2023, 1950, 1940

20: ^{1}H NMR (δ, toluene-d$_8$):   0.78 (d, 3 H)
2.94 (m, 1 H)
4.81 (t, 2 H)
6.02 (t, 2 H)
6.98 (t, 1 H)

partial IR (cm^{-1}): 2109, 2073, 1840

21: ^{1}H NMR (δ, C$_6$D$_6$):   0.55 (d, 3 H)
2.24 (m, 1 H)
2.43 (ddd, J = 6.5, 3, 1.5 Hz, 1 H)
2.98 (ddd, J = 6.5, 2, 1.5 Hz, 1 H)
3.22 (s, 3 H)
3.24 (s, 3 H)
3.29 (m, 2 H)
4.72 (dd, J = 6.5, 5 Hz, 1 H)
4.91 (dd, J = 6.5, 5 Hz, 1 H)

9. (cont.)

21: partial IR (cm^{-1}): 2041, 1993, 1752

^{13}C{^{1}H} NMR (δ, C$_6$D$_6$): 21.8
33.6
40.0
52.0
57.5
66.7
73.6
86.5
89.4
168.8
168.9
223.0

10. Determine the structure of the product of the following reaction.

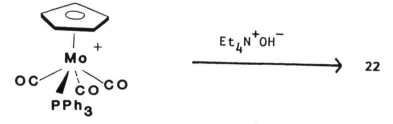

Et$_4$N$^+$OH$^-$

$\longrightarrow$ **22**

Spectral data for **22**:

IR (Nujol, cm^{-1}): 1955, 1862, 1616

^{1}H NMR (δ, CD$_2$Cl$_2$): 5.08 (d, J = 1.3 Hz, 5 H)
7.42 (m, 16 H)

^{13}C{^{1}H} NMR (δ, CD$_2$Cl$_2$): 95.5 (s)
128.9 (s)
129.0 (s)
130.9 (s)
133.2 (s)
209.5 (d, J = 11 Hz)
238.5 (d, J = 26 Hz)

11. The following reaction sequence has recently been observed:

Ph-I $\xrightarrow[\substack{\text{butadiene} \\ \text{Et}_2\text{NH}}]{\text{Pd(OAc)}_2, \text{ PPh}_3}$

Ph⌇⌇⌇⌇NEt$_2$

+

Ph⌇⌇⌇⌇

Propose a mechanism for the formation of the two products.

12. a) The following reaction gives exclusively one product. What product is formed and why are no others formed?

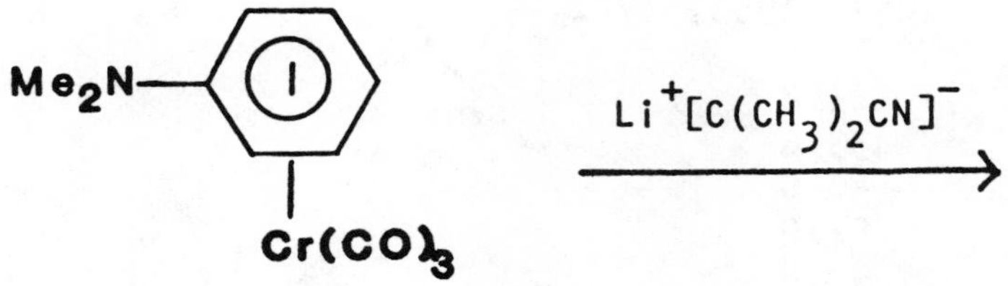

Me$_2$N—⬡(I) $\xrightarrow{\text{Li}^+[\text{C(CH}_3)_2\text{CN]}^-}$

Cr(CO)$_3$

b) The reaction shown below also gives exclusively one product; however, here the regiochemistry is <u>different</u> from that observed above. What is the product formed and why is a different regiochemistry observed?

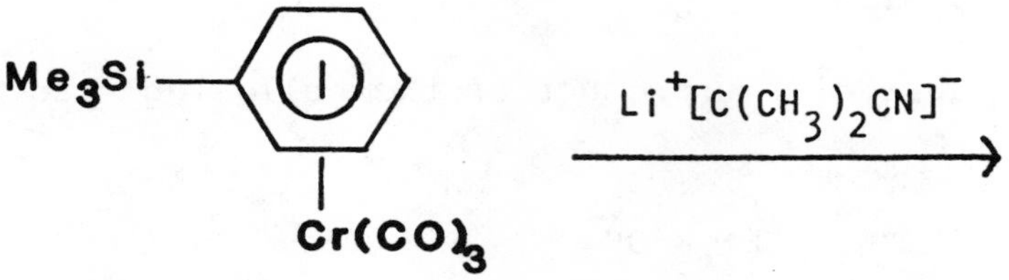

Me$_3$Si—⬡(I) $\xrightarrow{\text{Li}^+[\text{C(CH}_3)_2\text{CN]}^-}$

Cr(CO)$_3$

13. A variety of organic products can be obtained from the reaction of a palladium–olefin complex with nucleophilic reagents, as shown below. Note that addition of hydrogen to the reaction mixture gives the saturated alkyl product, and simple warming to room temperature gives the substituted olefin. Another interesting aspect of the reaction is that stabilized carbanions (pK_a < 15) usually attack at the more-substituted end of the olefin, whereas non-stabilized carbanions (pK_a > 15) are less selective and often give products that result from attack at the less-substituted end of the olefin.

R$_2$NH   Cl
    Pd
Cl   ‖
          R

$\xrightarrow[\substack{-78^\circ\text{C} \\ \text{THF}}]{\overset{\cdot\cdot}{\text{Nu}}}$

R$_2$NH   Cl
    Pd
Cl
      Nu   R

R.T.

H$_2$

Nu   R

Nu   R

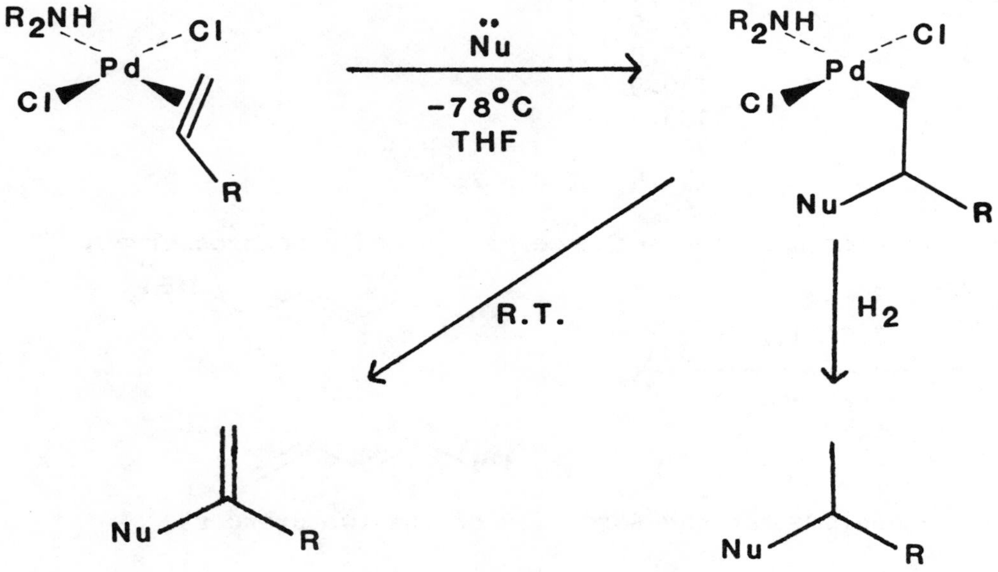

13. (cont.) Using this reaction sequence, design syntheses for the following organic molecules.

a)

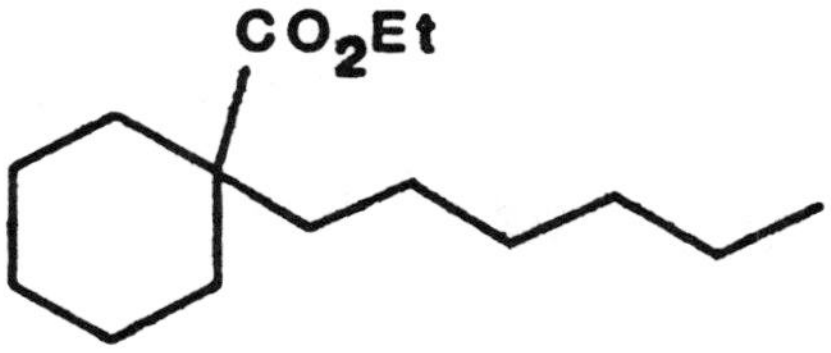

b)

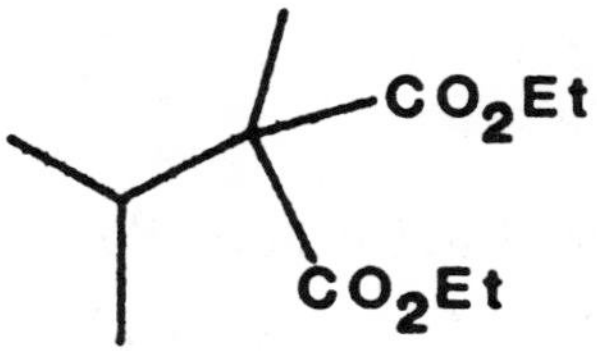

c)

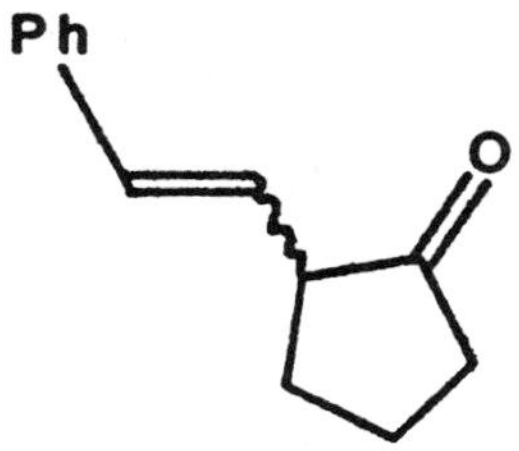

d)

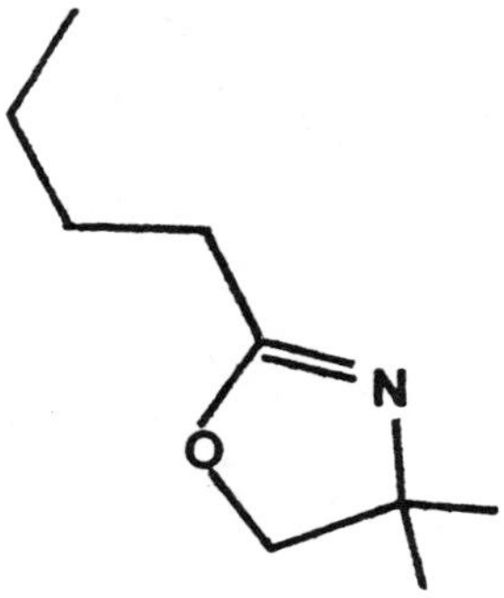

e)

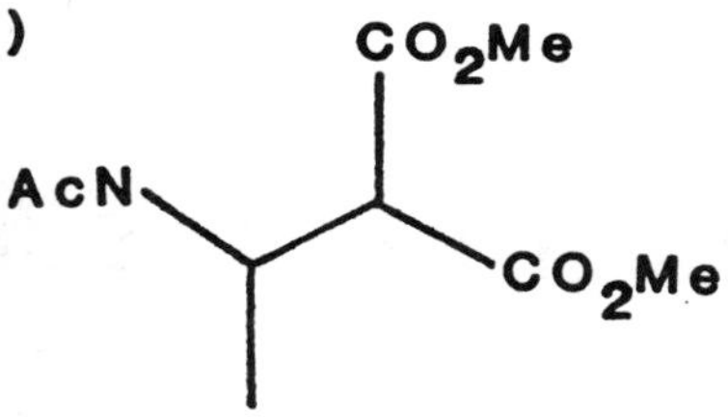

f)

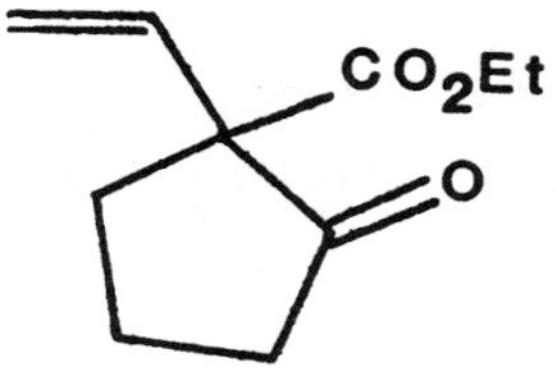

g)

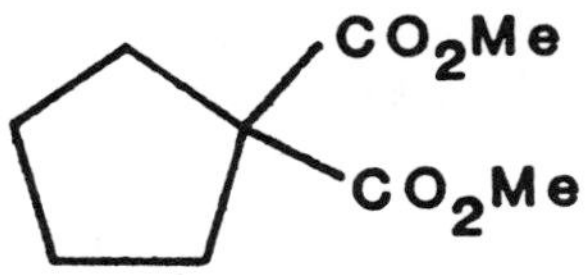

ANSWERS

1. a) 1 = CpRe(CO)(NO)(CHO)

Casey, C.P.; Andrews, M.A.; Rinz, J.E. *J*. *Am*. *Chem*. *Soc*. **1979**, <u>101</u>, 741.

b) 2 = CpRe(CO)(NO)(CO$_2$H)

Casey, C.P.; Andrews, M.A.; Rinz, J.E. *J*. *Am*. *Chem*. *Soc*. **1979**, <u>101</u>, 741.

c) 3 = (CO)$_5$Cr–C(OCH$_3$)(CH$_3$)(PR$_3$)

Fischer, E.O. <u>Adv</u>. <u>Organomet</u>. <u>Chem</u>. **1976**, <u>14</u>, 1.

d) 4 = Cl$_2$(AsPh$_3$)Pd=C(NH(p-tol))$_2$

Crociani, B.; Boschi, T.; Nicolini, M.; Belluco, U. <u>Inorg</u>. <u>Chem</u>. **1972**, <u>11</u>, 1292.

e)

5 =

Reger, D.L.; Belmore, K.A.; Mintz, E.; McElligott, P.J. <u>Organomet</u>. **1984**, <u>3</u>, 134.

f)

6 =

Semmelhack, M.F.; Hall, H.T.; Yoshifuji, M.; Clark, G. *J*. *Am*. *Chem*. *Soc*. **1975**, <u>97</u>, 1247.

2. Briefly, the rules state that unsaturated hydrocarbon ligands can be classified as **even** or **odd** (depending on their hapto (η) number), and **open** for acyclic (or nonconjugated cyclic) ligands or **closed** for cyclic conjugated systems. Under kinetically controlled conditions and for 18-electron systems, nucleophilic attack will occur preferentially at an **even** ligand before an **odd** one, and on an **open** system before a **closed** one. The preferential site of nucleophilic attack for **even open** polyenes will be at a terminal carbon atom (unless sterically hindered), and for **odd open** polyenes, no attack at the terminal positions is seen unless the metal is very electron-deficient.

2. a)

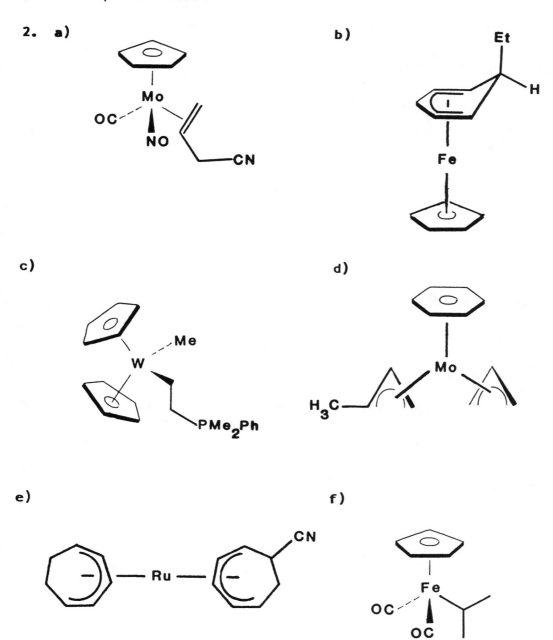

b)

c)

d)

e)

f)

Davies, S.G.; Green, M.L.H.; Mingos, D.M.P.   Tetrahedron, 1978, 34, 3047.

3.

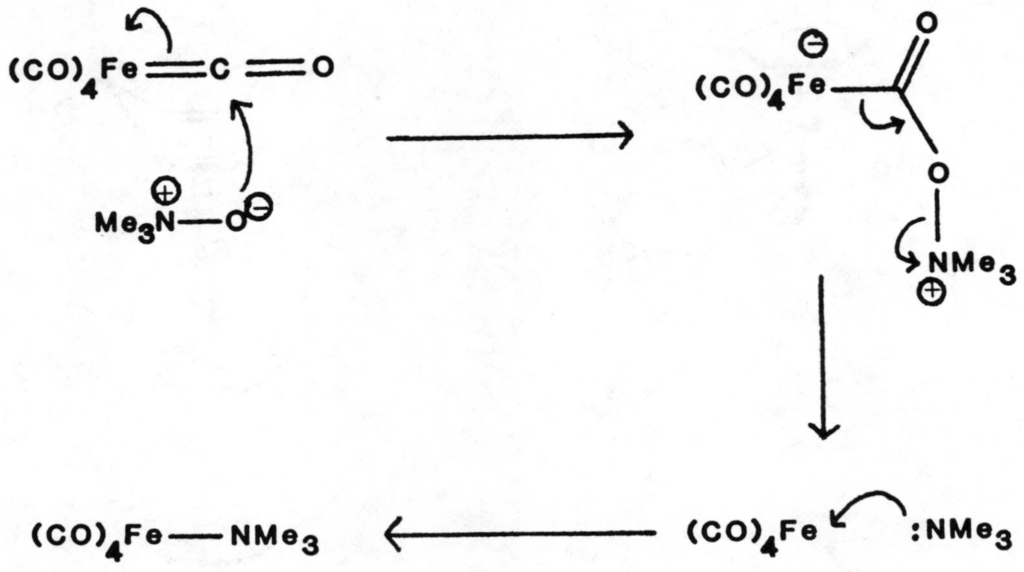

$(CO)_4Fe$=C=O

Me_3N—$O^{\ominus}$

$\longrightarrow$

$(CO)_4Fe^{\ominus}$—$C(=O)$—O—$NMe_3^{\oplus}$

$(CO)_4Fe$—NMe_3 $\longleftarrow$ $(CO)_4Fe$ $:NMe_3$

7

Elzinga, J.; Hogeveen, H. J. Chem. Soc. Chem. Comm., **1977**, 705.

4. In the scheme below, Pd(0) represents the transition metal and an unknown number of associated ligands.

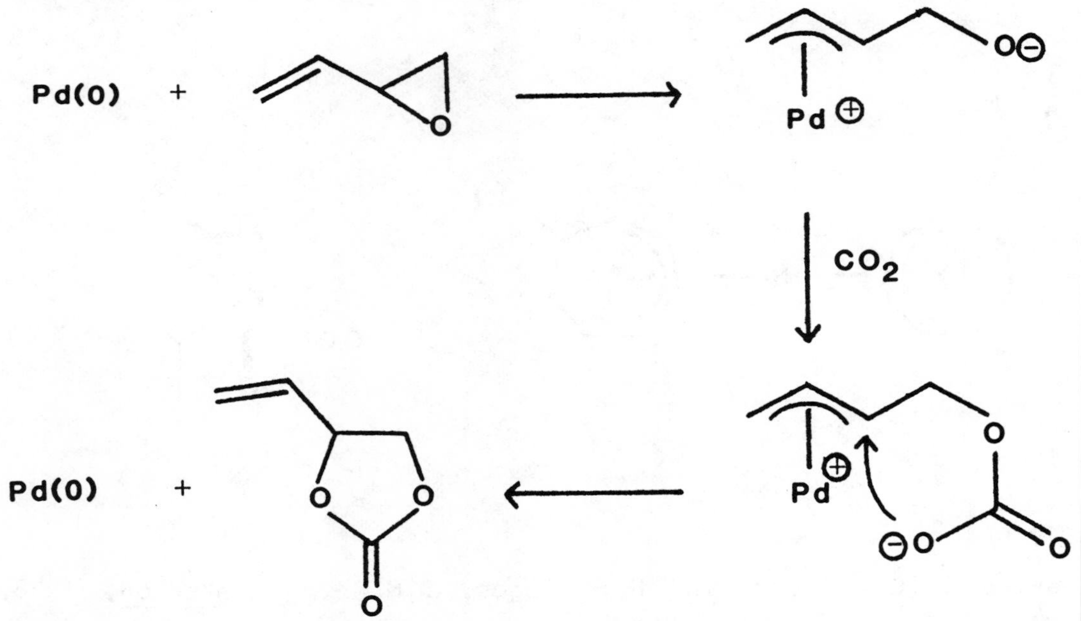

Pd(0) + [vinyl epoxide] $\longrightarrow$ [π-allyl Pd complex with alkoxide]

CO_2 $\downarrow$

Pd(0) + [vinyl cyclic carbonate] $\longleftarrow$ [π-allyl Pd carbonate complex]

Fujinami, T.; Suzuki, T.; Kamiya, M.; Fukuzawa, S.; Sakai, S. Chem. Lett., **1985**, 199.

5. a) The proposed mechanism is shown below.

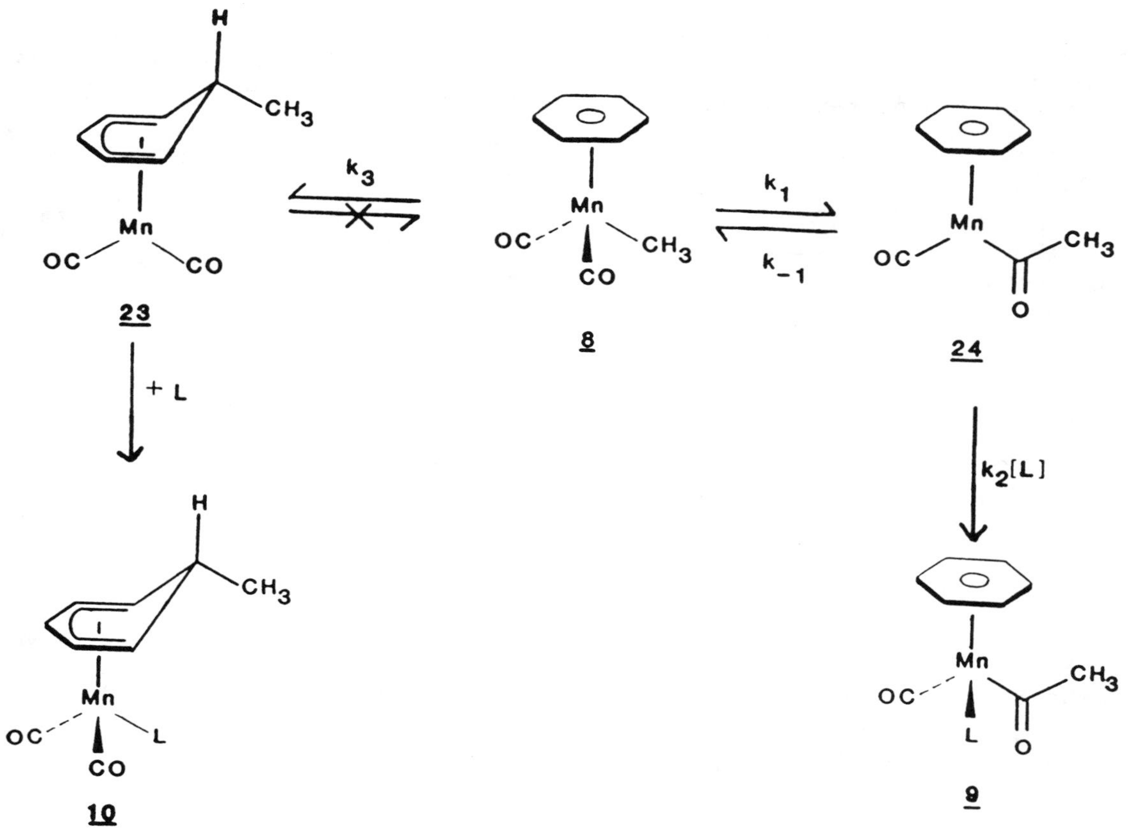

b) Since a carbon–carbon bond is very difficult to break, it is unlikely that the conversion of **8** to **23** is a reversible reaction. Assuming that **23** and **24** are steady-state intermediates, the following kinetic expressions can be obtained:

$$\frac{d[9]}{dt} = \frac{k_1 k_2 [8][L]}{k_{-1} + k_2[L]} \quad \text{and} \quad \frac{d[10]}{dt} = k_3[8]$$

c) At high concentrations of external ligand, the kinetic expression for formation of **9** reduces to :

$$d[9]/dt = k_1[8]$$

Since **9** is the observed product, k_1 must be larger than k_3, a very reasonable assumption. At low ligand concentrations, the kinetic expression for the formation of **9** becomes:

$$\frac{d[9]}{dt} = \frac{k_1 k_2 [8][L]}{k_{-1}}$$

The second-order reaction at small [L] is slow; therefore, the first-order formation of **10** predominates.

5. (cont.)

NOTE

The mechanism for the formation of **10** may be more complex than is indicated here. For the benzyl analog, the following mechanism has recently been proposed for migration of the benzyl group to the ring.

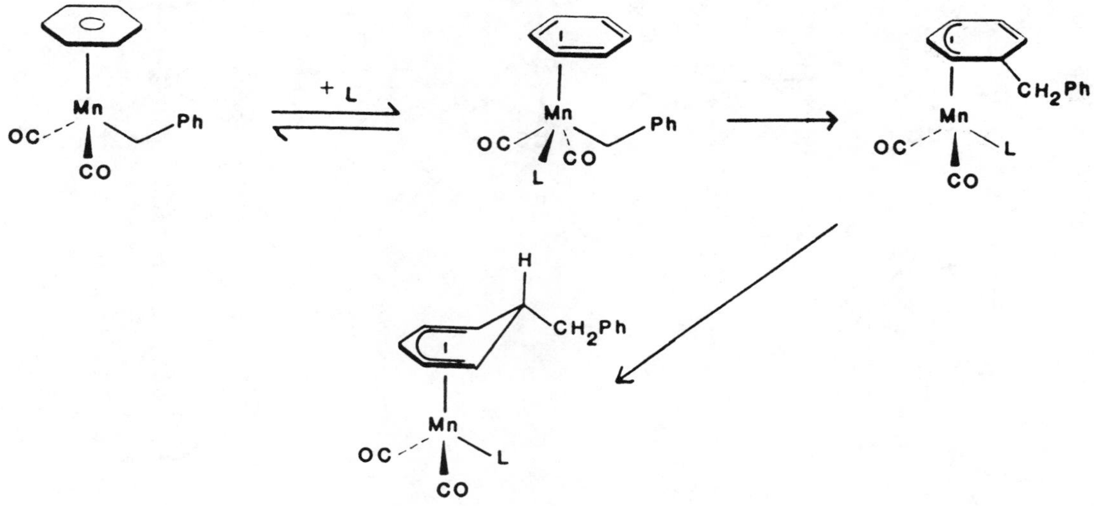

Brookhart, M.; Pinhas, A.R.; Lukacs, A. Organomet. **1982**, **1**, 1730.

Brookhart, M., personal communication.

6. a)

12 =

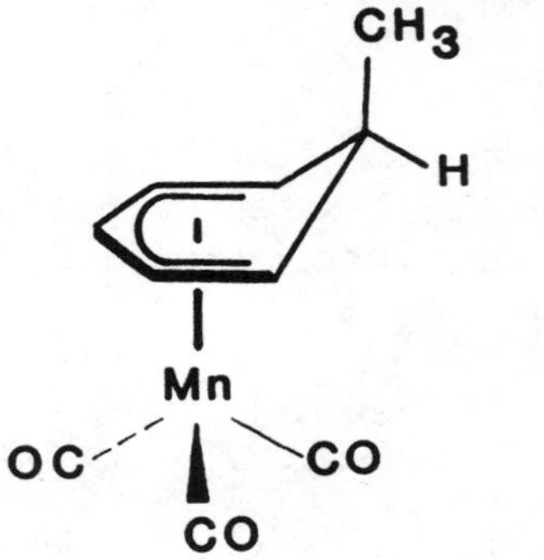

This product is formed by a direct nucleophilic attack on the coordinated benzene ring.

b)

13 =

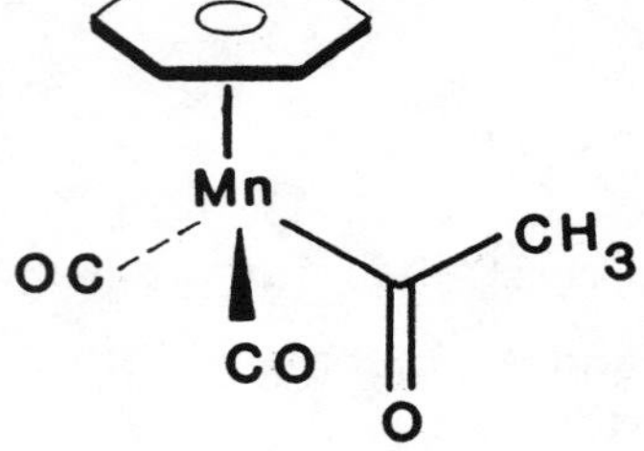

This product is formed by a nucleophilic attack on the carbonyl.

c)

14 =

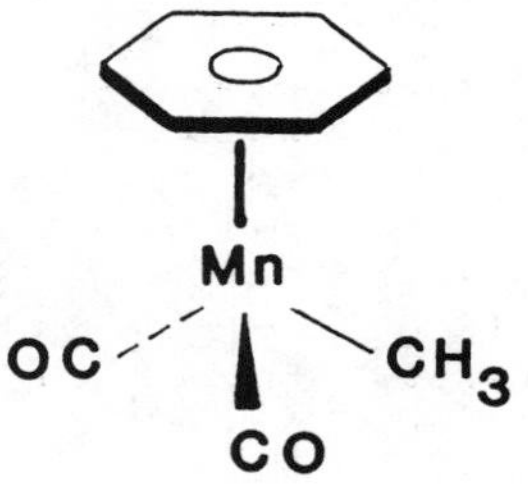

This product arises from **13**, but the manner in which decarbonylation occurs is not known.

6. (cont.) The reason for the different reactivities of MeLi and LiMe$_2$Cu with [(benzene)Mn(CO)$_3$]$^+$ is not known.

Reference for **a)**: Munro, G.A.M.; Pauson, P.L. <u>Isr. J. Chem.</u>, **1976–7**, <u>15</u>, 258.

Reference for **b)** and **c)**: Brookhart, M.; Pinhas, A.R.; Lukacs, A. <u>Organomet.</u>, **1982**, <u>1</u>, 1730.

7. a) For <u>cis</u> (internal) nucleophilic attack, the product will have the methyl groups <u>cis</u> to one another.

b) For <u>trans</u> (external) nucleophilic attack, the product will have the methyl groups <u>trans</u> to one another.

7. c) The value of the coupling constant J_{ab} is indicative of the dihedral angle ϕ between H_a and H_b, with $J_{ab} = 10\cos^2\phi$.

| | 15a | | 15b |

The relative magnitudes of the coupling constants J_{ab} for the products expected from the reaction of the amine with both cis- and trans-2-butene by the different modes of attack are listed in the table below.

Olefin	Expected J_{ab}		Observed J_{ab}
	Internal attack	External attack	
cis-2-butene	large	small	2 Hz
trans-2-butene	small	large	6 Hz

The NMR data agree with the external or trans mode of nucleophilic attack, thus the product of the reaction of cis-2-butene with benzylamine is 15a and the product of the reaction of trans-2-butene with benzylamine is 15b.

Wong, P.K.; Madhavaroa, M.; Marten, D.F.; Rosenblum, M. J. Am. Chem. Soc. 1977, 99, 2823.

8.

8. (cont.)

18 =

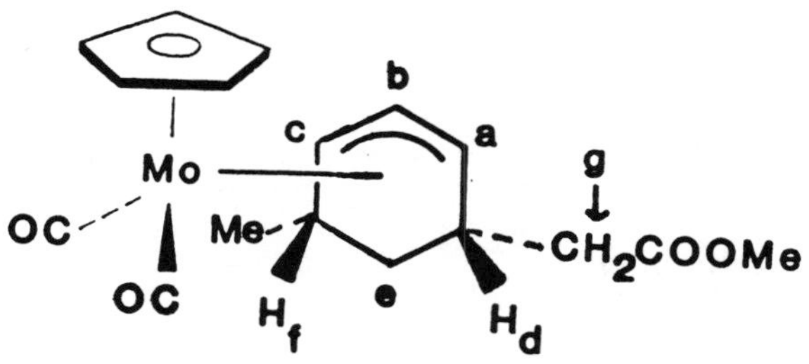

Assignments for spectral data for 18:

IR: Peaks at 1945 and 1870 cm^{-1} are due to metal carbonyl stretches.
     The peak at 1740 cm^{-1} is due to the ester carbonyl.

^{1}H NMR (δ, CDCl$_3$):   0.65 (d, J = 14.5 Hz, 1 H, exo H$_e$)
                         0.85 (dm, J = 14.5 Hz, 1 H, endo H$_e$)
                         1.22 (d, J = 7.2 Hz, 3 H, CH$_3$)
                         1.90 (m, 1 H, H$_f$)
                         2.30 (m, 1 H, H$_d$)
                         2.55 (d, J = 6.8 Hz, 2 H, H$_g$)
                         3.5-3.75 (m, 2 H, H$_a$, H$_c$)
                         3.69 (s, 3 H, -COOCH$_3$)
                         4.17 (t, J = 7.0 Hz, 1 H, H$_b$)
                         5.29 (s, 5 H, Cp)

Pearson, A.J.; Khan, M.N.I. J. Am. Chem. Soc. 1984, 106, 1872.

9.

19 =

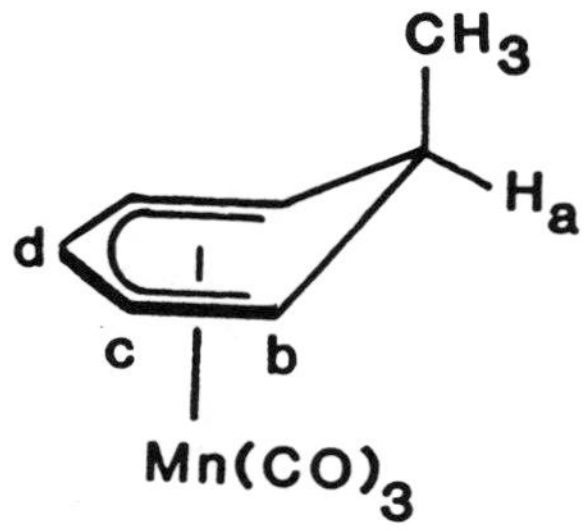

Assignments for spectral data for 19:

IR: All peaks given are metal carbonyl stretches.

9. (cont.) Spectral data for **19**:

^{1}H NMR (δ, CDCl$_3$):   0.4 (d, 3 H, CH$_3$)
                        2.5 (m, 1 H, H$_a$)
                        3.1 (t, 2 H, H$_b$)
                        4.6 (t, 2 H, H$_c$)
                        5.6 (tt, 1 H, H$_d$)

20 =

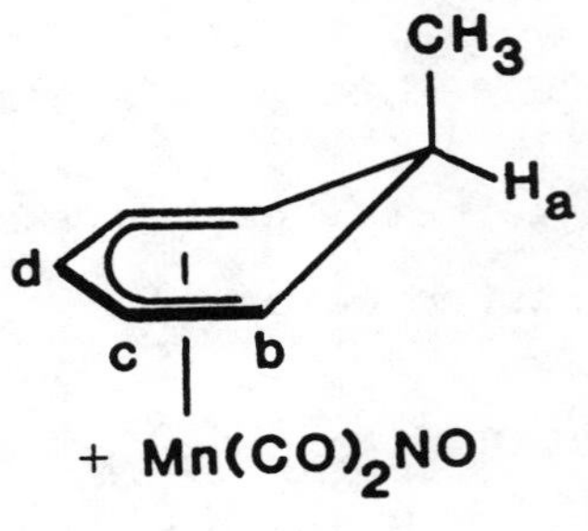

Assignments for spectral data for **20**:

IR: Peaks at 2109 and 2073 cm^{-1} are metal carbonyl stretches.
     Peak at 1840 cm^{-1} is an N–O stretch of the nitrosyl.

^{1}H NMR (δ, toluene-d$_8$):   0.78 (d, 3 H, CH$_3$)
                           2.94 (m, 1 H, H$_a$)
                           4.81 (t, 2 H, H$_b$)
                           6.02 (t, 2 H, H$_c$)
                           6.98 (t, 1 H, H$_d$)

21 =

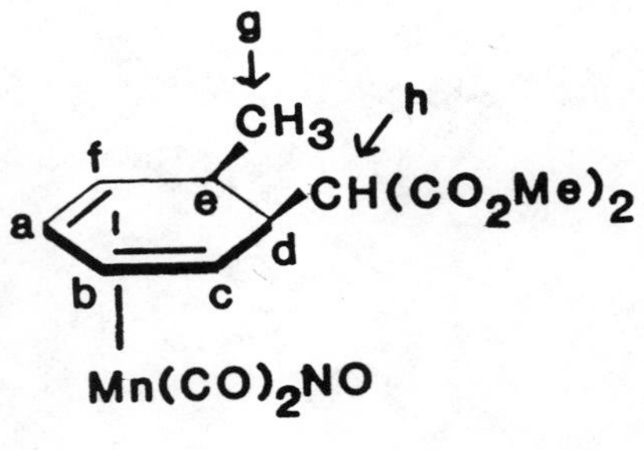

Assignments for spectral data for **21**:

IR: Peaks at 2041 and 1993 cm^{-1} are metal carbonyl stretches.
     Peak at 1752 cm^{-1} is an N–O stretch of the nitrosyl.

9. (cont.) Spectral data for **21**:

^{1}H NMR (δ, C_6D_6): 0.55 (d, 3 H, $CH_3(g)$)

2.24 (m, 1 H, H_e)

2.43 (ddd, J = 6.5, 3, 1.5 Hz, 1 H, H_f)

2.98 (ddd, J = 6.5, 2, 1.5 Hz, 1 H, H_c)

3.22 (s, 3 H, $-CO_2CH_3$)

3.24 (s, 3 H, $-CO_2CH_3$)

3.29 (m, 2 H, H_d, H_h)

4.72 (dd, J = 6.5, 5 Hz, 1 H, H_a)

4.91 (dd, J = 6.5, 5 Hz, 1 H, H_b)

^{13}C{^{1}H} NMR (δ, C_6D_6): 21.8 (C_g)

33.6 (C_e)

40.0 (C_d)

52.0 ($CO_2\underline{C}H_3$)

57.5 (C_h)

66.7 (C_f)

73.6 (C_c)

86.5 (C_a)

89.4 (C_b)

168.8 ($\underline{C}O_2CH_3$)

168.9 ($\underline{C}O_2CH_3$)

223.0 (Mn–CO)

Chung, Y.K.; Sweigart, D.A.; Connelly, N.G.; Sheridan, J.B. <u>J</u>. <u>Am</u>. <u>Chem</u>. <u>Soc</u>. **1985**, <u>107</u>, 2388.

Spectra of **19**: Munro, G.A.M.; Pauson, P.L. <u>Isr</u>. <u>J</u>. <u>Chem</u>. **1976–7** , <u>15</u>, 258.

10.

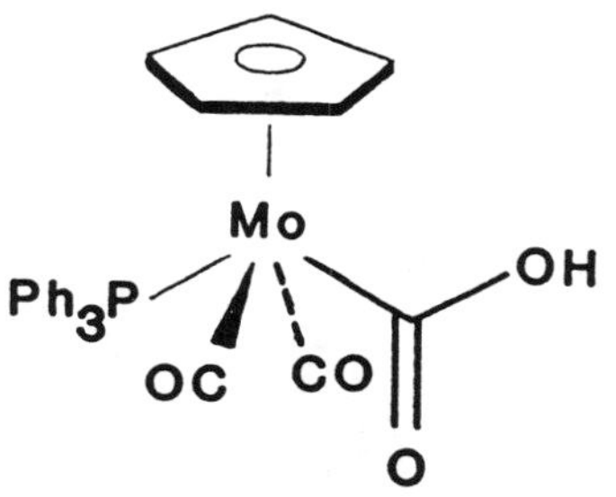

10. (cont.)

IR(cm^{-1})	^{13}C NMR	^1H NMR
1616 acyl C–O stretch	95.5 (Cp)	5.08 (Cp)
1955 and 1862 M–CO stretches	128–134 (Ph carbons)	7.42 (Ph and
	209.5 (COOH)	COOH)
	238.5 (M–CO)	

The PPh$_3$ and COOH ligands must be trans for the carbonyls to be equivalent in the ^{13}C NMR spectrum.

Gibson, D.H.; Owens, K.; Ong, T.-S. J. Am. Chem. Soc., 1984, 106, 1125.

11. Under the reaction conditions, the Pd(II) is reduced to Pd(0) by the amine. The Pd(0) species then undergoes an oxidative-addition with phenyl iodide to regenerate a Pd(II) complex.

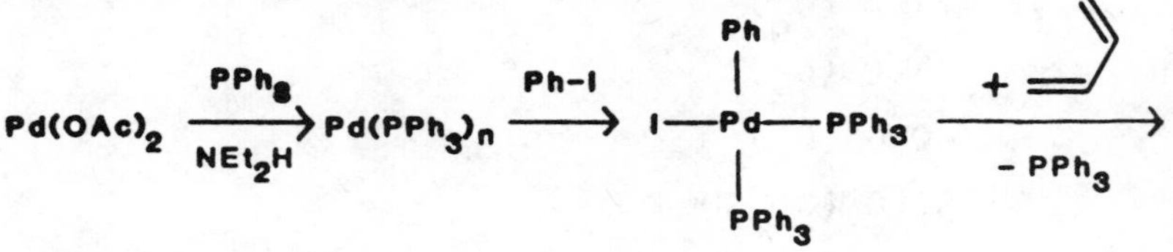

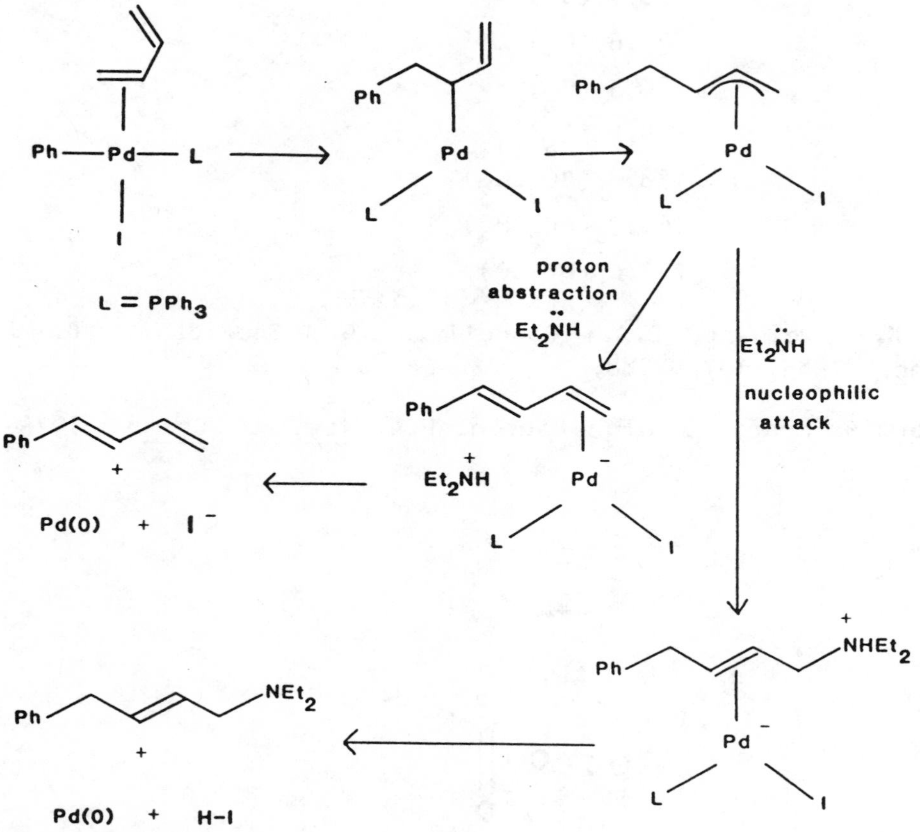

$$Pd(OAc)_2 \xrightarrow[NEt_2H]{PPh_3} Pd(PPh_3)_n \xrightarrow{Ph-I} I-Pd-PPh_3 \xrightarrow{+ \quad , \quad -PPh_3}$$

L = PPh$_3$

proton abstraction
Et$_2$NH

Et$_2$NH
nucleophilic attack

Pd(0) + I$^-$

Pd(0) + H–I

O'Connor, J.M.; Stallman, B.J.; Clark, W.G.; Shu A.Y.L.; Spada, R.E.; Stevenson, T.M.; Dieck, H.A. J. Org. Chem., 1983, 48, 807.

12. a) and **b)** The two products are:

It has been found that these reactions are subject to both orbital control and to charge control. The following conclusions about the position of nucleophilic attack have been drawn.

(1) Steric effects are important in comparing primary, secondary, and tertiary carbanion nucleophiles, with ortho substitution nearly completely inhibited with tertiary carbanions.

(2) When there are no steric effects present, the amount of ortho substitution becomes comparable to the amount of meta substitution, with the ratio reaching one to one with reactive anions. Usually, there is more meta substitution than ortho.

(3) Para substitution is only important when large alkyl substituents are present on the arene ring.

(4) With more stable nucleophiles, more para substitution is observed.

Since the nucleophile is tertiary, no ortho substitution is expected or observed; thus, it must be either meta or para subsitution. According to the rules, the anion should attack the aniline complex at the meta position, which is what is observed. This is due to orbital control.

Since the two reactions have different regiochemistries and since ortho attack has been ruled out, the trimethylsilyl complex must be attacked at the para position. A possible explanation is that large substituents direct the attack to the para position, as predicted by rule (3).

Another explanation for the different regiochemistry observed for the two reactions is based on the effect the donor amine and the acceptor silyl group have on the lowest unoccupied orbital (LUMO) of the arenechromium tricarbonyl complex, a top view of which is shown on the next page. In benzenechromium tricarbonyl, the LUMO is a degenerate set of two orbitals; however, upon substitution, this degeneracy is broken, as shown on the following page.

12. (cont.)

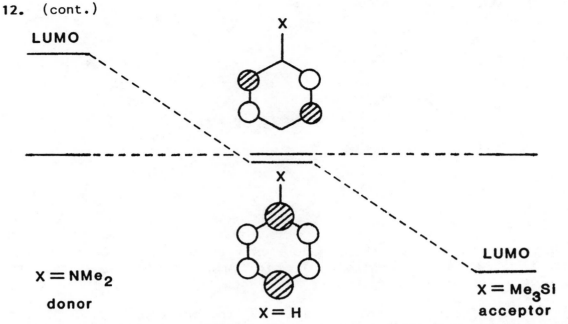

X = NMe₂

donor

X = H

X = Me₃Si

acceptor

The LUMO of the aniline complex has non-zero coefficients at the ortho and meta positions, and for reasons stated above, meta attack is preferred. On the other hand, the silyl-substituted complex has its largest coefficient in the LUMO at the para position, and thus, para substitution is observed.

Semmelhack, M.F.; Hall, H.T.; Farina, R.; Yoshifuji, M.; Clark, G.; Bargar, T.; Hirotsu, K.; Clardy, J. J. Am. Chem. Soc. 1979, 101, 3535.

Semmelhack, M.F.; Clark, G.R.; Farina, R.; Saeman, M. J. Am. Chem. Soc., 1979, 101, 217.

Albright, T.A.; Carpenter, B.K. Inorg. Chem. 1980, 19, 3092.

Semmelhack, M.F.; Garcia, J.L.; Cortes, D.; Farina, R.; Hong, R.; Carpenter, B.K. Organomet. 1983, 2, 467.

13. In all cases, Pd = $(R_2NH)PdCl_2$

a)

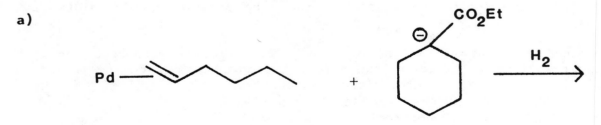

13. b)

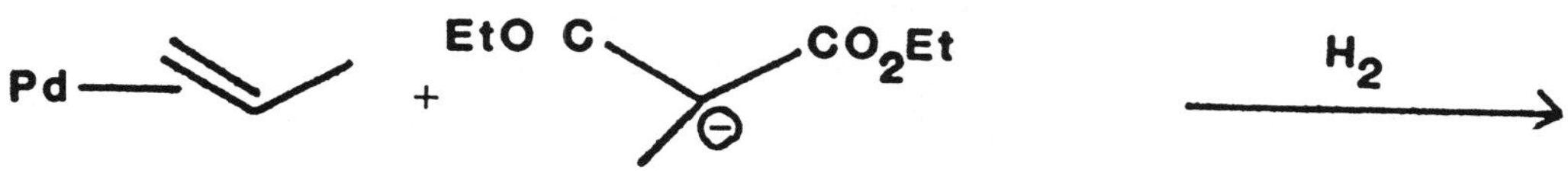

c)

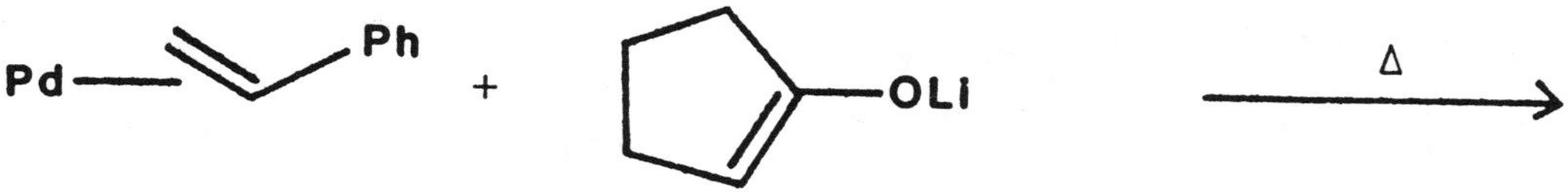

d)

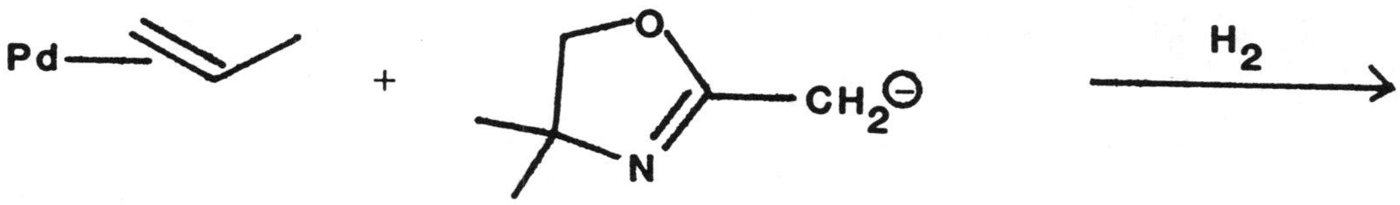

e)

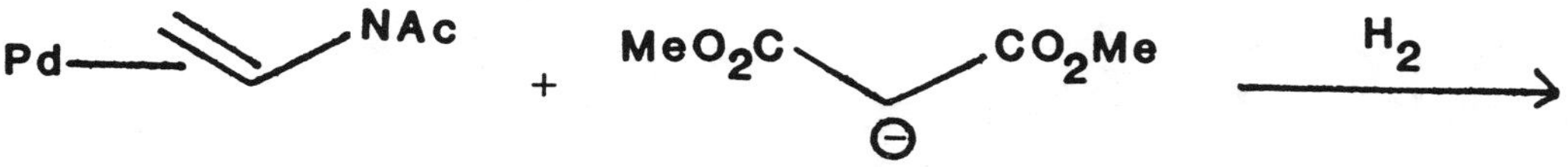

f)

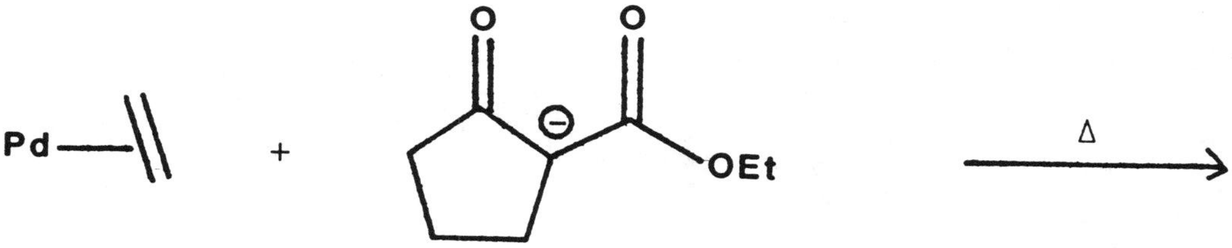

g)

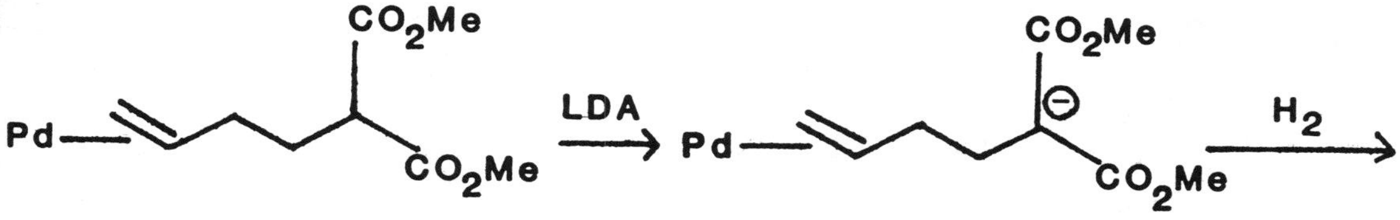

Hegedus, L.S.; Williams, R.E.; McGuire, M.A.; Hayashi, T. _J_. _Am_. _Chem_. _Soc_. **1980**, _102_, 4973.

Electrophilic Attack on Coordinated Ligands

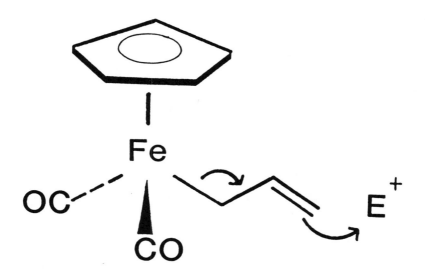

QUESTIONS

1. Propose a structure for the product of each of the following reactions. Give a mechanism for the formation of each product.

a)

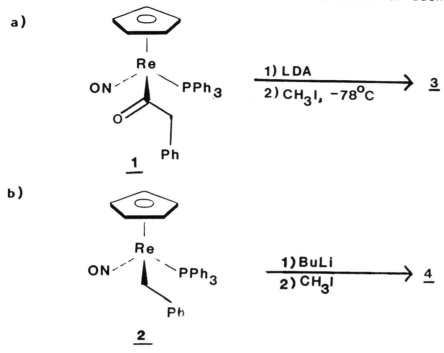

1) LDA
2) CH₃I, −78°C → **3**

1

b)

1) BuLi
2) CH₃I → **4**

2

Spectral data:

For **3**

^{1}H NMR (δ, CDCl₃)

7.47–7.21 (m, 20 H)
5.48–4.12 (br m, 4 H)
3.78 (s, 2 H)
0.82 (d, J=5.5 Hz, 3 H)

^{13}C NMR (δ, CD₂Cl₂)
(broad-band decoupled)

194.1 (s)
135.9 (s)
135.4 (d, J=53.2 Hz)
134.2 (d, J=10.5 Hz)
131.1 (s)
130.2 (s)
129.1 (d, J=11.3 Hz)
129.0 (s)

For **4**

^{1}H NMR (δ, CDCl₃)

8.03–7.37 (m, 20 H)
5.84–4.79 (m, br, 4 H)
4.11 (dd, J=11.5, 8.5 Hz, 1 H)
3.26 (dd, J=11.5, 2.0 Hz, 1 H)
2.36 (s, 3 H)

^{13}C NMR (δ, CD₂Cl₂)
(broad-band decoupled)

159.9 (d, J=3.5 Hz)
137.0 (d, J=50.8 Hz)
134.3 (d, J=15.1 Hz)
130.6 (s)
129.0 (d, J=10.8 Hz)
128.0 (s)
127.7 (s)
122.3 (s)

(^{13}C data continued on the following page)

1. (cont.)

Spectral data (cont.):

For 3	For 4
^{13}C NMR (δ, CD_2Cl_2)	^{13}C NMR (δ, CD_2Cl_2)
(broad-band decoupled)	(broad-band decoupled)

127.3 (s)	105.3 (s)
103.0 (s)	95.5 (d, J=4.0 Hz)
96.1 (s)	92.3 (s)
95.0 (d, J=5.7 Hz)	89.6 (s)
86.3 (s)	86.3 (s)
85.1 (s)	13.4 (s)
53.8 (s)	−2.0 (d, J=4.2 Hz)
−30.0 (d, J=6.9 Hz)	

IR (cm^{-1}, CH_2Cl_2)	IR (cm^{-1}, thin film)
1660(m), 1640(s)	1628(s)

2. The iron compounds **5a** and **5b** suffer Fe–C bond cleavage by bromine, as shown.

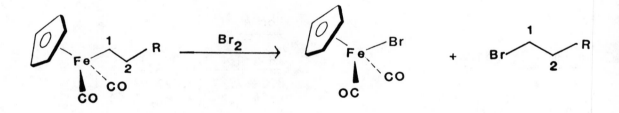

5a: R = t-Bu
5b: R = Ph

a) What experiment can be done to determine the stereochemistry of this reaction at carbon 1?

b) The results of such an experiment are that compound **5a** undergoes cleavage with inversion of configuration at carbon 1, whereas compound **5b** undergoes cleavage with retention of configuration at carbon 1. Suggest an explanation for these results.

c) Design an experiment to test your hypothesis for part **b)**.

3. Give structures for compounds **7–14**.

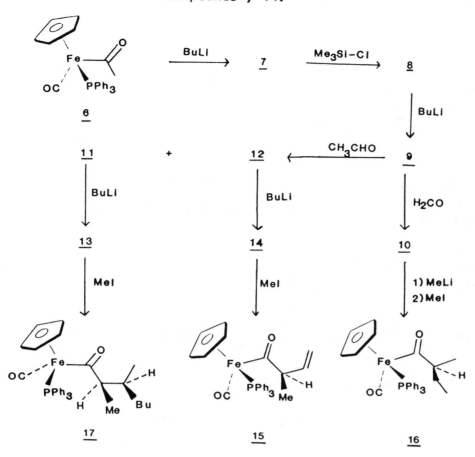

4. Give the product and propose a mechanism for the following transformation:

Spectral data for **18**:

^{1}H NMR (δ, CS$_2$)
4.67–4.90 (m, 4 H)
0.31 (s, 9 H)
0.18 (s, 3 H)

IR (cm^{-1}, in CHCl$_3$)
1991, 1937

5. The following reaction occurs, with products **19** and **20** obtained in a 1:3 ratio.

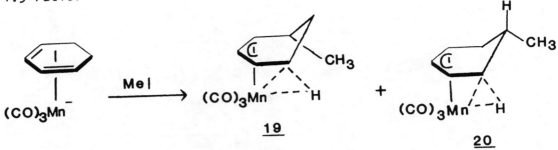

19   **20**

a) Suggest a reason for the observed product ratio.

b) Propose a mechanism for the formation of each product.

6. Provide structures for the products of the following reactions (Fp = CpFe(CO)$_2$).

a)   Fp⌇⌇⌇   +   Me$_3$O$^+$BF$_4^-$ ——————————→   **22**

21

b)   **21**   +   (CN)$_2$C=C(CN)$_2$ ——————————→   **23**

c)   **21**   +   [Fp(ethylene)]$^+$ ——————————→   **24**

7. Propose structures for compounds **25** and **26**.

1) (CH$_3$)$_3$SiOTf, −78°C
—————————————————————→   **25**
2) Fp⌇⌇⌇ , −78°C

25   ——————————→   C$_{21}$H$_{18}$O$_6$Fe$_2$, a bicyclic ketone
1) 40°C
2) H$_2$O workup   **26**

8.   Identify the products of the following reactions:

$[CpMo(CO)_3)]_2$ + $AgBF_4$ $\xrightarrow[\text{Ph}]{\text{Ph}—\!\!\equiv\!\!—}$ **28** + **29**
   27

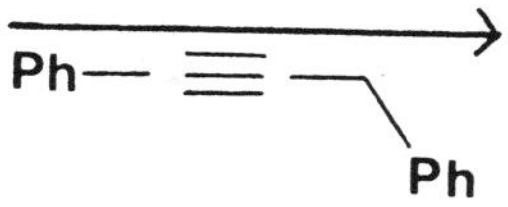

27   + Cl_2(1 eq) $\longrightarrow$ **30**

27   +   excess Cl_2 $\longrightarrow$ **31**

HINT
Compounds **28** and **29** are isomeric.

Spectral data:

Compound	^{1}H NMR (δ, $CDCl_3$)	IR (cm^{-1})
27	5.29 (s)	2010, 1945, 1910
28	7.6–6.6 (m, 20 H) 5.90 (s, 5 H) 5.10 (d, J=16.6 Hz, 1 H) 5.09 (d, J=17.5 Hz, 1 H) 4.72 (d, J=16.6 Hz, 1 H) 4.71 (d, J=17.5 Hz, 1 H)	2050
29	7.6–6.6 (m, 20 H) 6.03 (s, 5 H) 4.79 (d, J=17.1 Hz, 2 H) 4.32 (d, J=17.1 Hz, 2 H)	2050
30	4.25 (s)	2055, 1983, 1960
31	6.51 (s, acetone-d_6)	2105, 2063

ANSWERS

1.   a) Initial deprotonation of the Cp ring of **1** is followed by migration of the acyl group to the ring. Reaction with methyl iodide then serves to alkylate the metal to give **3**.

b) Because the alkyl group of **2** does <u>not</u> migrate to the ring, alkylation by methyl iodide occurs at the ring to give product **4**.

Assignments for spectral data:

For 3   For 4

^{1}H NMR (δ, CDCl$_3$) ^{1}H NMR (δ, CDCl$_3$)

7.47–7.21 (m, 20 H, Ph's)   8.03–7.37 (m, 20 H, Ph's)
5.48–4.12 (br m, 4 H, C$_5$H$_4$)   5.84–4.79 (m, br, 4 H, C$_5$H$_4$)

3.78 (s, 2 H, CH$_2$Ph)   4.11 (dd, J=11.5, 8.5 Hz, 1 H,
0.82 (d, J=5.5 Hz, 3 H, –CH$_3$)   –CHH'Ph)
                                          3.26 (dd, J=11.5, 2.0 Hz, 1 H,
                                               –CHH'Ph)
                                          2.36 (s, 3 H, –CH$_3$)

1. (cont.) Assignments for spectral data:

<u>For 3</u>

^{13}C NMR (δ, CD$_2$Cl$_2$)

(broad-band decoupled)
___

194.1 (s, C=O)
135.9 (s, ipso-C of acyl Ph)
135.4 (d, J=53.2 Hz, ipso-C
    of PPh$_3$)
134.2 (d, J=10.5 Hz, C

    of PPh$_3$)
131.1 (s, p-C of PPh$_3$)
130.2 (s, C of acyl Ph)
129.1 (d, J=11.3 Hz, C of PPh$_3$)
129.0 (s, C of acyl Ph)
127.3 (s, C of acyl Ph)
103.0 (s, C$_5$H$_4$, ipso-C

    of C$_5$H$_4$)
96.1 (s, C$_5$H$_4$)
95.0 (d, J=5.7 Hz, C$_5$H$_4$)
86.3 (s, C$_5$H$_4$)
85.1 (s, C$_5$H$_4$)
53.8 (s, CH$_2$Ph)
-30.0 (d, J=6.9 Hz, CH$_3$)

<u>For 4</u>

^{13}C NMR (δ, CD$_2$Cl$_2$)

(broad-band decoupled)
___

159.9 (d, J=3.5 Hz, ipso-C of
    benzyl group)
137.0 (d, J=50.8 Hz, ipso-C
    of PPh$_3$)
134.3 (d, J=15.1 Hz, C of PPh$_3$)
130.6 (s, p-C of PPh$_3$)
129.0 (d, J=10.8 Hz, C of PPh$_3$)
128.0 (s, C of benzyl group)
127.7 (s, C of benzyl group)
122.3 (s, C of benzyl group)
105.3 (s, ipso-C of C$_5$H$_4$)
95.5 (d, J=4.0 Hz, C$_5$H$_4$)
92.3 (s, C$_5$H$_4$)
89.6 (s, C$_5$H$_4$)
86.3 (s, C$_5$H$_4$)
13.4 (s, CH$_3$)
-2.0 (d, J=4.2 Hz, Re-$\underline{C}$H$_2$Ph)

Heah, P.C.; Gladysz, J.A. <u>J</u>. <u>Am</u>. <u>Chem</u>. <u>Soc</u>. **1984**, <u>106</u>, 7636.

2. a) In order to determine the stereochemistry of the reaction, the experiment should be carried out with the labeled compounds **5c** and **5d**. If threo/erythro interchange occurs, the reaction occurs with inversion of configuration; otherwise, the reaction occurs with retention of configuration. Experimentally, this is verifiable by NMR. See answer to problem 2, Chapter 14 for details of this analysis.

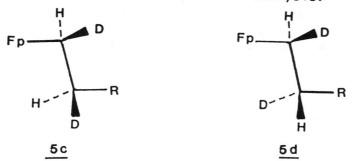

      <u>5c</u>   **<u>5d</u>**

2. **b)** For R = t-Bu, **5a**, a normal S_N2 attack by bromide occurs with inversion at carbon 1.

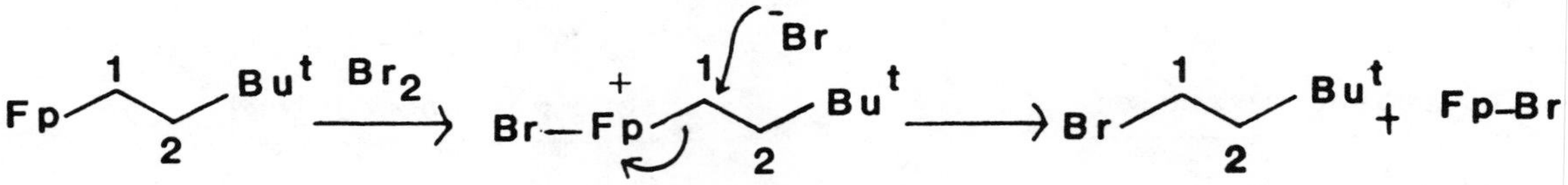

5a

For R = Ph, **5b**, a carbocation is formed and stabilized by neighboring phenyl participation. Attack by bromide on this carbocation results in a net retention of configuration at carbon 1.

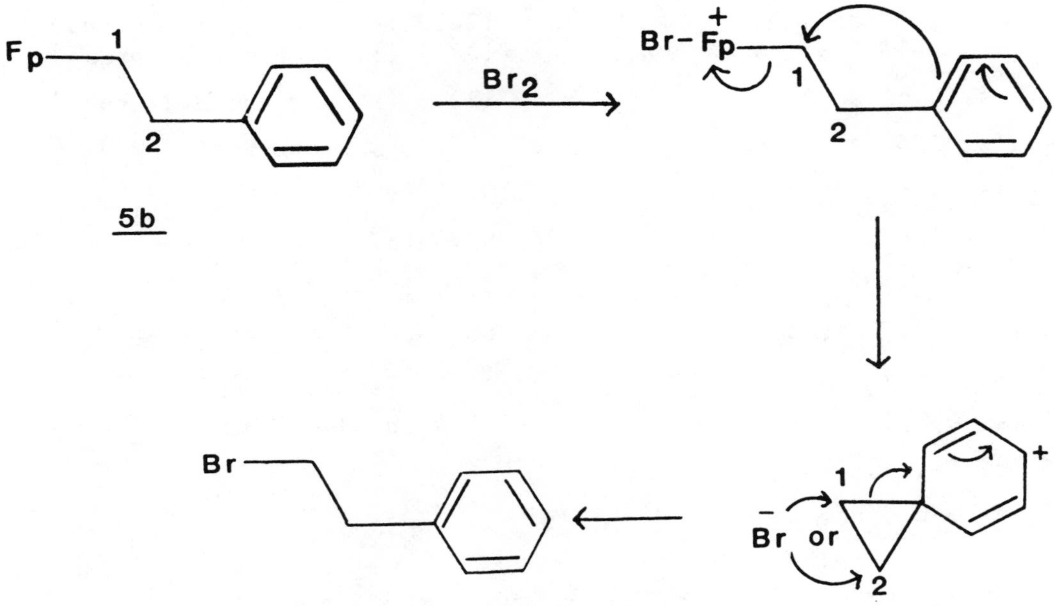

5b

c) To verify this experimental hypothesis, the experiment can be carried out using 1,1-dideuterio-**5a** and **5b**. The t-butyl compound, **5a**, should give exclusively 1,1-dideuteriobromoalkane and the phenyl compound, **5b**, should give a 50:50 mixture of 1,1-dideuterio- and 2,2-dideuteriobromoalkanes (excluding any isotope effect).

Whitesides, G.M.; Boschetto, D.J. J. Am. Chem. Soc. **1971**, <u>93</u>, 1529.

2. The reactions and products are detailed below.

Fpp=CpFe(CO)(PPh₃)

Note that these alkylation reactions are diastereoselective, with the products arising from attack on only one face of the enolate or acyl compounds. Davies and Seeman have postulated a set of rules that govern the stereochemistry of the formation of chiral centers at the organic ligand in complexes of the type $CpM(PPh_3)(L)R$, where M = Fe, Co, Mn, Re; L = CO, NO; R = prochiral fragment.[3] The basis for these rules is that one phenyl ring of the PPh_3 seems to prefer to be in a plane parallel to that of the $L-M-C_\alpha$ entity, and any conformations of the R group that result in positioning of an alkyl or aryl substituent in between the $C_\alpha-M-L$ plane and the PPh_3 plane are energetically unfavorable.

Additionally, the acyl oxygen prefers to be _anti_ to the CO, i.e., the
$O-C_\alpha-M-CO$ dihedral angle is $180°$. Thus **12** is simply deprotonated by
Bu-Li to give **14**, while **11** undergoes a nucleophilic attack by Bu-Li to
give **13**.

Davies, S.G.; Walker, J.C. _J. Chem. Soc., Chem. Comm._ **1985**, 209.

Davies, S.G.; Seeman, J.I. _Tetrahedron Let._ **1984**, 1845.

4.

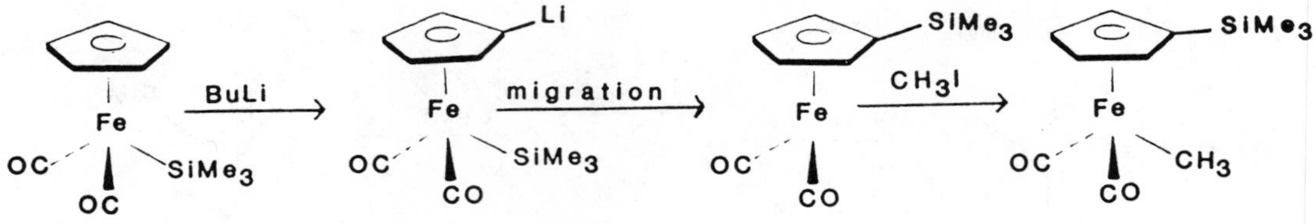

Assignments for NMR data for **18**:

^{1}H NMR (δ, CS_2):

4.67-4.90 (m, 4 H, Cp)
0.31 (s, 9 H, Me_3Si-)
0.18 (s, 3 H, CH_3)

Berryhill, S.R.; Sharenow, B. _J. Organomet. Chem._, **1981**, _221_, 143.

5.   a) The product ratio is probably due to steric reasons. The major
isomer has the methyl group farther away from the $Mn(CO)_3$.

b) A mechanism for the formation of product **19** is shown below:

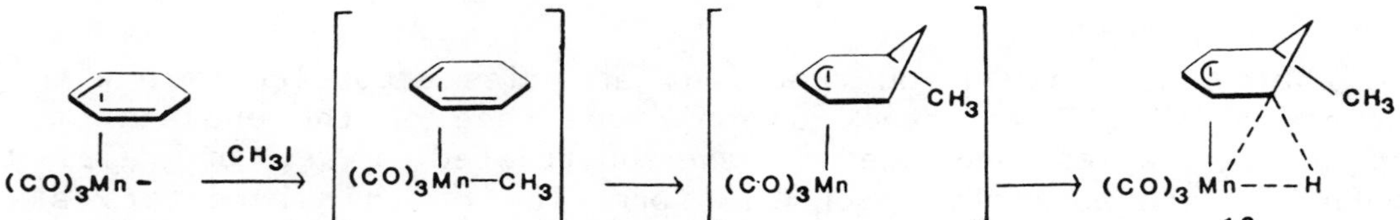

A mechanism for the formation of product **20** is shown below:

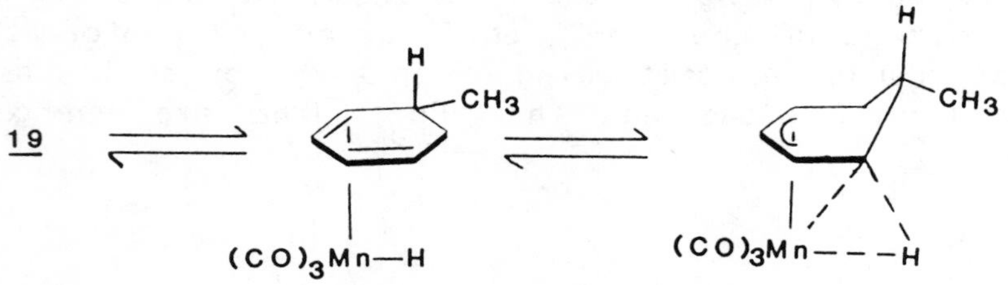

Brookhart, M.; Lamanna W.; Pinhas, A.R. _Organomet._, **1983**, _2_, 638.

6. a)

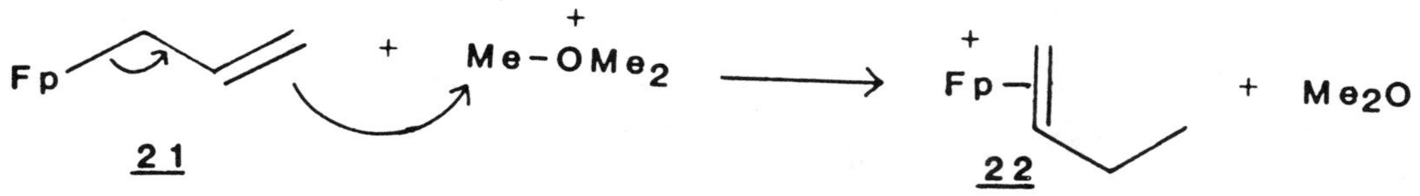
Fp
+
Me−OMe2
21
Fp
+
Me2O
22

b)

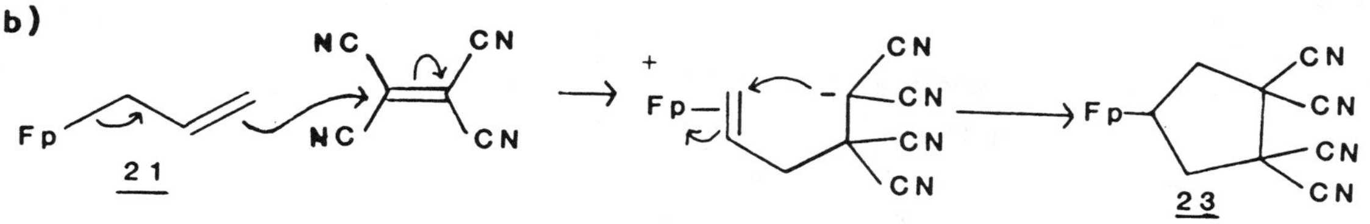
NC
CN
Fp
NC
CN
21
Fp
CN
CN
CN
CN
Fp
CN
CN
CN
CN
23

c)

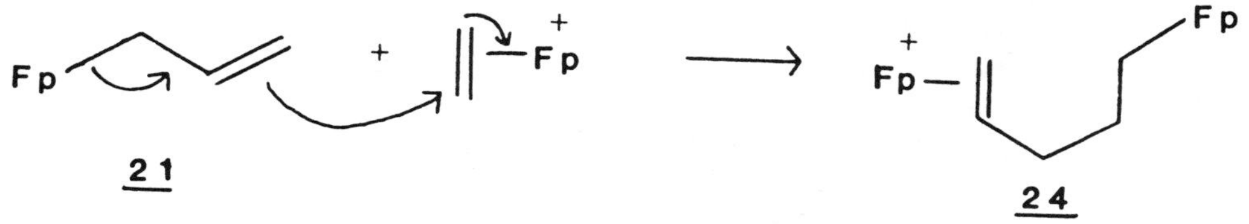
Fp
+
Fp
21
Fp
Fp
24

Rosenblum, M. Accts. Chem. Res. 1974, 7, 122.

7.

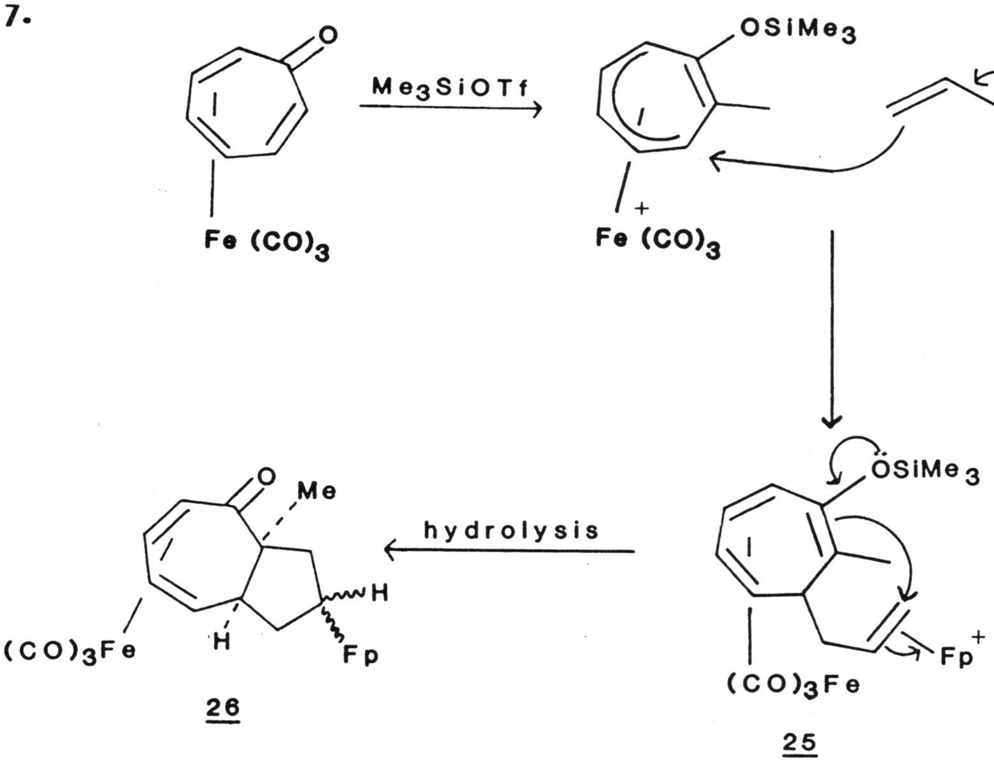
O
Me3SiOTf
Fe (CO)3
OSiMe3
Fe (CO)3
Fp
OSiMe3
(CO)3Fe
Fp
25
hydrolysis
O
Me
H
(CO)3Fe
H
Fp
26

Watkins, J.C.; Rosenblum, M. Tetrahedron Let. 1985, 26, 3531.

8. Compounds **28–31** are shown below:

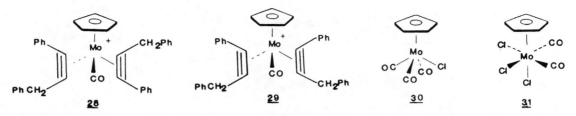

The NMR data are assigned as follows:

Compound	^{1}H NMR (δ, $CDCl_3$)

28
7.6–6.6 (m, 20 H, Ph's)
5.90 (s, 5 H, Cp)
5.10 (d, J=16.6 Hz, 1 H, CHH'Ph)
5.09 (d, J=17.5 Hz, 1 H, CHH'Ph)
4.72 (d, J=16.6 Hz, 1 H, CHH'Ph)
4.71 (d, J=17.5 Hz, 1 H, CHH'Ph)

29
7.6–6.6 (m, 20 H, Ph's)
6.03 (s, 5 H, Cp)
4.79 (d, J=17.1 Hz, 2 H, CHH'Ph)
4.32 (d, J=17.1 Hz, 2 H, CHH'Ph)

30
4.25 (s, Cp)

31
6.51 (s, acetone-d_6, Cp)

NOTE
In these systems, low field Cp resonances and high energy CO stretches are characteristic of electron deficient or cationic species.

For **28** and **29**: Allen, S.R.; Beevor, R.G.; Green, M.; Norman, N.C.; Orpen, A.G.; Williams, I.D. J. Chem. Soc., Dalt. Trans. **1985**, 435.

For **30** and **31**: Burckett-St. Laurent, J.C.T.R.; Field, J.S.; Haines, R.J.; McMahon, M. J. Organomet. Chem. **1979**, 181, 117.

Haines, R.J.; Nyholm, R.S.; Stiddard, M.H.B. J. Chem. Soc. (A) **1966**, 1606.

9

Metallacycles

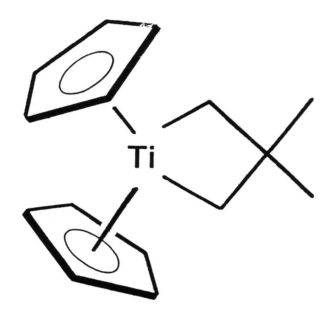

QUESTIONS

1. Predict the products of the following reactions:

a) Cp_2TiCl_2 +

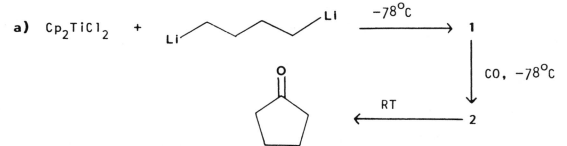

b) $Cp^*_2Ti(C_2H_4)$ + C_2H_4 ⟶ 3

^{1}H NMR data for 3 (δ, toluene-d_8):

1.80 (s, 15 H)
1.72 (m, 4 H)
0.60 (m, 4 H)

c)

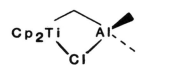 + $\xrightarrow[0°C]{DMAP}$ 4

HINT
See problem 6 in this chapter (DMAP = 4-dimethylaminopyridine).

d)

$Cp_2Mo\overset{+}{-}$ $\xrightarrow{NaBH_4}$ 5

^{1}H NMR data for 5 (δ, C_6D_6):

4.19 (s, 10 H)
3.55 (q, J=8 Hz, 2 H)
0.69 (t, J=8 Hz, 4 H)

e)

Cp^*_2Th + 2 △ $\xrightarrow[cyclohexane]{60°C}$ 6

1. e) (cont.)

^{1}H NMR data for **6** (δ, ppm):	^{13}C NMR data for **6** (δ, ppm):
0.78 (d, 4 H)	123.47
−0.03 (d, 4 H)	70.33
−0.72 (m, 2 H)	11.54
	8.59

f)

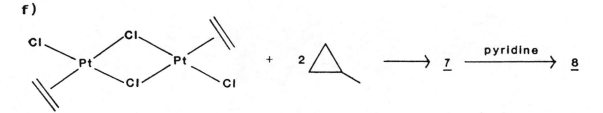

HINTS
1) Complex **7** is oligomeric and very insoluble in non-donor solvents.
2) Complex **8** exists as a mixture of two isomers.

2. The following reaction occurs to produce the novel metallacycle **9**.

$$(CO)_5Mn-\overset{\overset{O}{\|}}{C}-\overset{\overset{O}{\|}}{C}-Ph \quad \xrightarrow[\text{2) } PPN^+Cl^-]{\text{1) } LiEt_3BH, \text{ THF}} \quad 9$$

Spectral data for **9**:

^{1}H NMR (δ, acetone-d_6)

7.60, 7.23 (m, 35 H)
4.57 (s, 1 H)

IR (cm^{-1}, THF)

2043, 1959, 1940,
1664, 1636

^{13}C NMR (δ, CDCl$_3$)
(broad-band decoupled)

288.1 (s)
229.0 (s)
218.9 (s)
215.4 (s)
138.4 (s)
134.2 (s)
132.4 (t, J=6 Hz)
129.9 (t, J=7 Hz)
127.8 (s)
127.0 (d, J=108 Hz)
126.7 (s)
126.1 (s)
94.2 (s)

a) Provide a structure for **9** and propose a mechanism for its formation.

HINT
When the reaction is monitored by ^{1}H NMR, no low field resonances (below 8 ppm) are observed.

2. b) Complex **9** undergoes the transformations shown below.

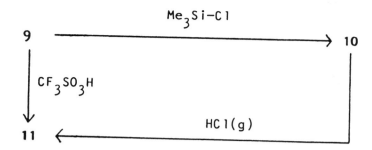

Spectral data:

For **10**

^{1}H NMR (δ, CCl_4)

7.33 (m, 5 H)
4.68 (s, 1 H)
0.20 (s, 9 H)

IR (cm^{-1}, $CHCl_3$)

2119, 2048, 2020, 1638

For **11**

^{1}H NMR (δ, $CDCl_3$)

7.2 (m, 5 H)
6.2 (s, 1 H)
5.0 (s, 1 H)

IR (cm^{-1}, $CHCl_3$)

3610–3260, 2116, 2052, 2016, 1639

Analytical data for **10**: 47.59% C, 3.91% H, 13.49% Mn

3. The following reactions have been observed.

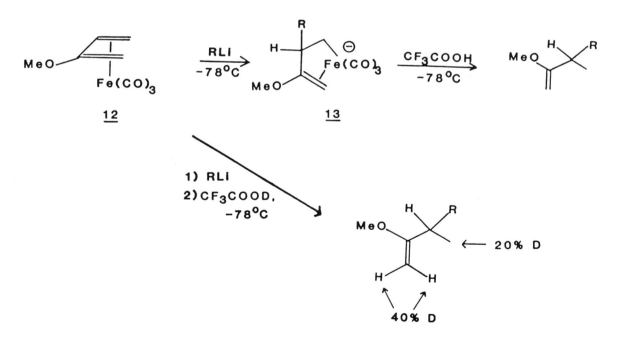

3. (cont.) Provide a mechanism which accounts for the deuterium scrambling in the products. Your mechanism does not necessarily have to account for the percentage of deuterium at each position, but only for the fact that deuterium does occur at three sites.

4. Propose reasonable structures for compounds **17**, **18**, **19**, and **20**. In all cases, L = PPh$_3$.

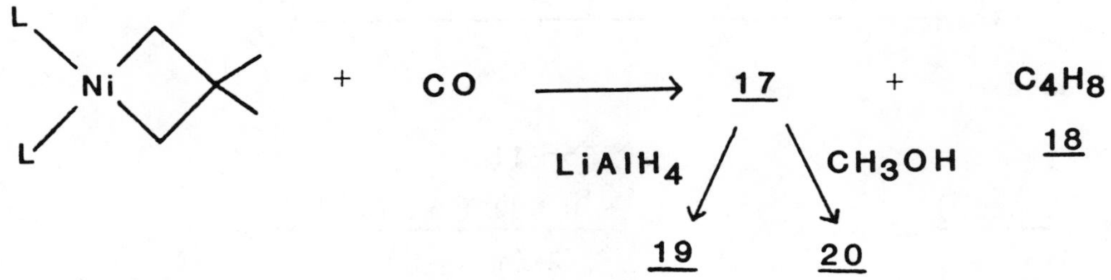

Spectral data for **17** (an organometallic complex):

^{1}H NMR (δ, acetone-d$_6$, -48°C)

2.35 (m, 2H)
7.45 (m, 30H)

^{13}C NMR (δ, toluene-d$_8$, -46°C)

38
176
aromatic C's

partial IR (cm^{-1}): 1750 (s)

Compound **18**: an alkene

Compound **19**: an alcohol

Compound **20**: an ester

5. Propose a mechanism for the formation of **21** and **22**.

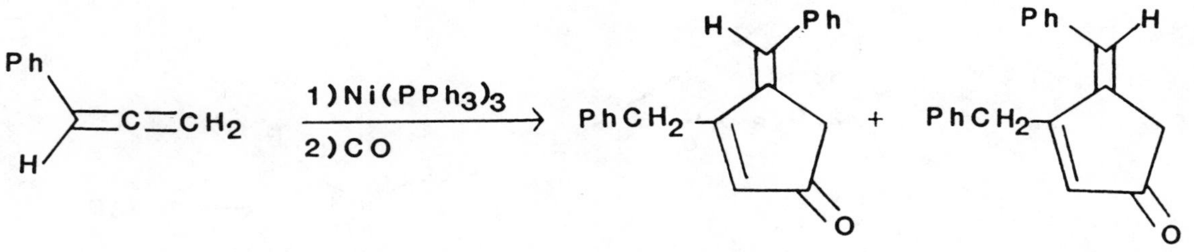

6. The following reactions are observed (DMAP = 4-dimethylamino-pyridine).

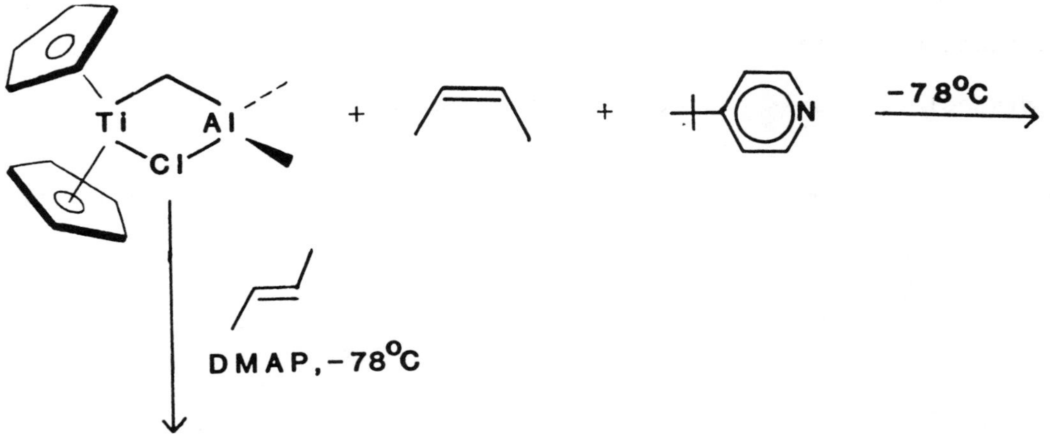

^{1}H NMR data (δ, toluene-d$_8$, -30°C):

For 23	For 24
0.03 (m, 1 H)	-0.80 (m, 1 H)
1.02 (d, 3 H)	1.04 (d, 3 H)
1.58 (d, 3 H)	1.68 (d, 3 H)
2.31 (t, 1 H)	2.33 (dd, 1 H)
3.51 (m, 2 H)	2.81 (dd, 1 H)
5.23 (s, 5 H)	3.05 (dd, 1 H)
5.32 (s, 5 H)	5.27 (s, 5 H)
	5.36 (s, 5 H)

a) Propose structures for compounds **23** and **24**.

b) What is the purpose of the amine in the reactions?

7. The following conversion has been shown to occur.

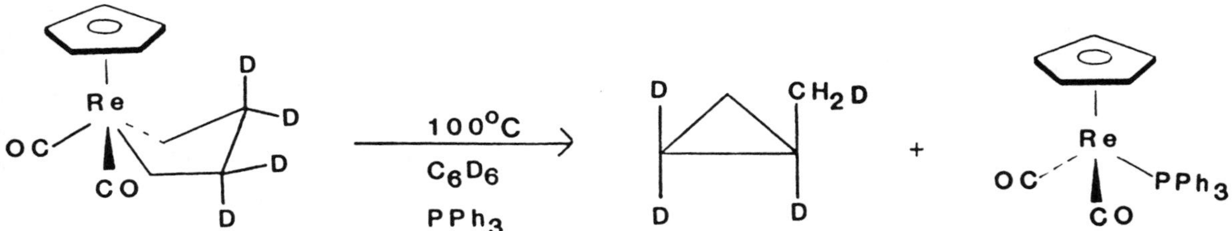

7. (cont.) Propose a mechanism consistent with the following observations:

1) The reaction is first order in **25**;

2) the reaction is zero order in phosphine; and

3) CO dissociation does <u>not</u> occur during the reaction.

8. Titanacyclobutane **26** is known to isomerize to titanacyclobutane **27**.

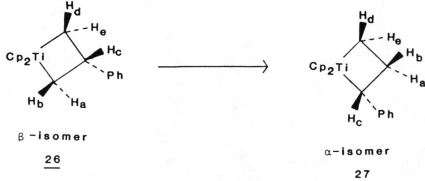

β-isomer

26

α-isomer

27

^{1}H NMR data (δ, C$_6$D$_6$):

β-isomer

6.99–7.40 (m, 5 H)
5.52 (s, 5 H)
5.40 (s, 5 H)
2.84 (t, J=9 Hz, 1 H)
2.39 (m, 2 H)
0.33 (m, 2 H)

α-isomer

6.71–7.35 (m, 5 H)
5.31 (s, 5 H)
5.08 (s, 5 H)
4.74 (dd, J=11, 9 Hz, 2 H)
3.12 (dd, J=9, 9 Hz, 2 H)
0.24 (m, 1 H)

You suspect that this rearrangement proceeds by the mechanism shown below. What experiments could be done to prove this mechanistic hypothesis?

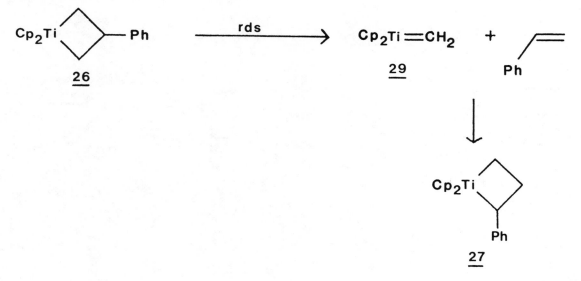

9. Propose a structure for product **31** and assign all spectroscopic data.

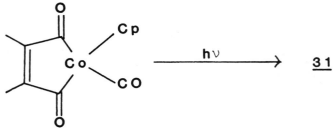

Spectral data for **31**:

^{1}H NMR (δ, ppm)	^{13}C NMR (δ, ppm)
1.91 (s, 6 H)	10.8
5.02 (s, 5 H)	51.3
	86.8
	225.9

partial IR (cm^{-1}): 1810, 1760

10. Thermolysis of **32** in cyclohexane at 150°C gives metallacycle **33**.

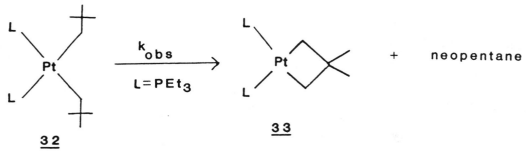

OBSERVATIONS

1) When excess PEt$_3$ is added, k$_{obs}$ decreases.

2) Thermolysis of **32** in cyclohexane-d$_{12}$ gives neopentane, C(Me)$_4$, that is 97% d$_0$ and 3% d$_1$.

3) Thermolysis of L$_2$Pt(CD$_2$–CMe$_3$)$_2$ in cyclohexane-d$_{12}$ gives neopentane that is 93% d$_2$ and 7% d$_3$.

Propose a mechanism for metallacycle formation that is consistent with the above observations.

ANSWERS

1. a)

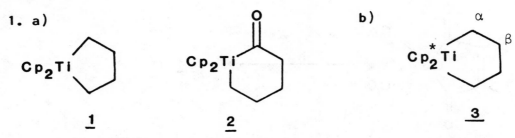

1   **2**   **3**

Assignments for ^{1}H NMR data for 3:

1.80 (s, 15 H, Cp*)
1.72 (m, 4 H, α-hydrogens of metallacycle)
0.60 (m, 4 H, β-hydrogens of metallacycle)

c)   **d)**

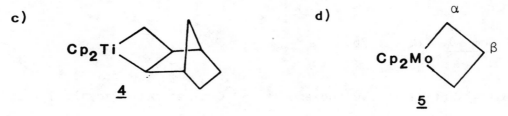

4   **5**

Assignments for ^{1}H NMR data for 5:

4.19 (s, 10 H, Cp)
3.55 (q, J=8 Hz, 2 H, β-hydrogens of metallacycle)
0.69 (t, J=8 Hz, 4 H, α-hydrogens of metallacycle)

e)   **f) 7 = [PtCl$_2$(C$_3$H$_5$Me)]$_n$**

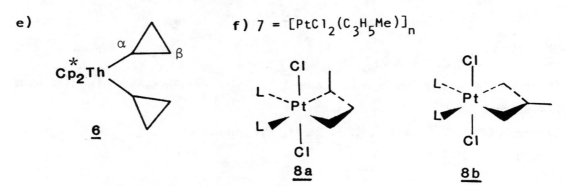

6   **8a**   **8b**

Assignments for spectral data for 6:

^{1}H NMR (δ, ppm)	^{13}C NMR (δ, ppm)
0.78 (d, 4 H, β-H's)	123.47 (C$_5$Me$_5$)
−0.03 (d, 4 H, β-H's)	70.33 (α-C's)
−0.72 (m, 2 H, α-H's)	11.54 (C$_5$Me$_5$)
	8.59 (β-C's)

1. (cont.) For a): McDermott, J.X.; Whitesides, G.M. J. Am. Chem. Soc. **1974**, _96_, 947.

For b): Cohen, S.A.; Auburn, P.R.; Bercaw, J.E. J. Am. Chem. Soc. **1983**, _105_, 1136.

For c): Gilliom, L.R.; Grubbs, R.H. J. Am. Chem. Soc. **1986**, _108_, 733.

For d): Ephritikhine, M.; Francis, B.R.; Green, M.L.H.; Mackenzie, R.E.; Smith, M.J. J. Chem. Soc., Dalt. Trans. **1977**, 1131.

For e): Fendrick, C.M.; Marks, T.J. J. Am. Chem. Soc. **1986**, _108_, 425.

For f): Al-Essa, R.J.; Puddephatt, R.J.; Perkins, D.C.L.; Rendle, M.C.; Tipper, F.H. J. Chem. Soc., Dalt. Trans. **1981**, 1738.

2. **a)** The proposed mechanism for the formation of **9** is shown below:

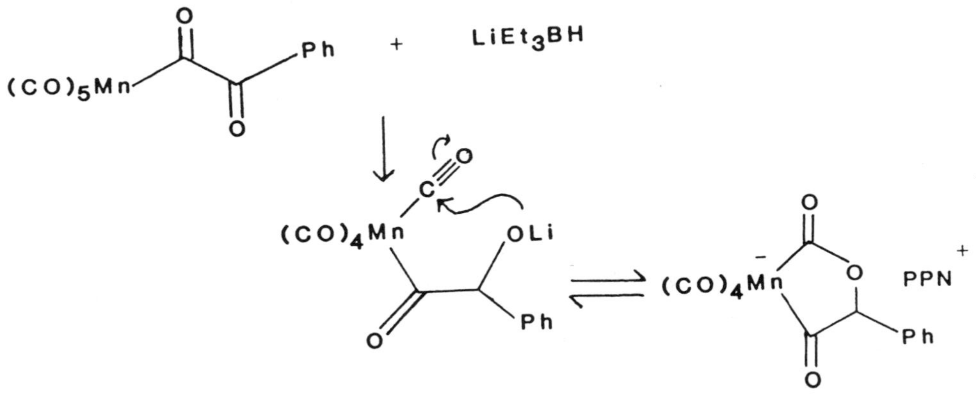

The fact that no low field resonances were observed during the formation of **9** indicates that hydride probably does not attack the carbonyl bound to the metal.

Assignments for spectral data for **9**:

^{1}H NMR (δ , acetone-d$_6$)

7.60, 7.23 (m, 35 H, PPN$^+$ and Ph)
4.57 (s, 1 H, C_HPh)

IR (cm^{-1}, THF)

M–C≡O stretches: 2043, 1959, 1940
acyl C=O stretches: 1664, 1636

^{13}C NMR (δ , CDCl$_3$)
(broad-band decoupled)

288.1, 229.0 (s, C=O)
218.9, 215.4 (s, M–CO)
138.4 (s, ipso-C of Ph)
134.2 (s, C of PPN$^+$)
132.4 (t, J=6 Hz, C of PPN$^+$)
129.9 (t, J=7 Hz, C of PPN$^+$)
127.8, 126.7, 126.1 (s, C of Ph)
127.0 (d, J=108 Hz, ipso-C of PPN$^+$)
94.2 (s, C_HPh)

2. b) Complexes **10** and **11** are shown below:

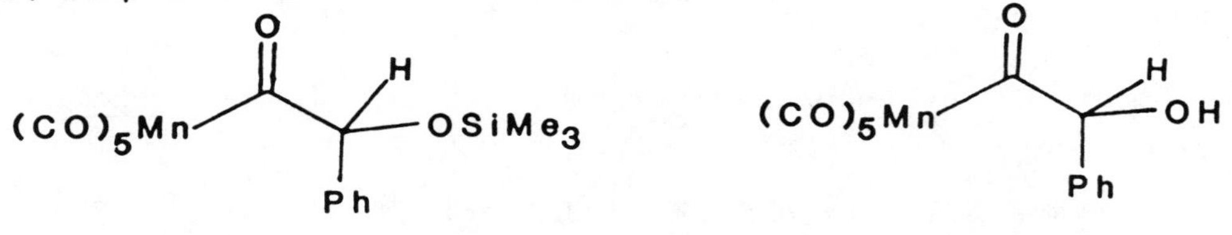

Assignments for spectral data:

For **10**

^{1}H NMR (δ , CCl$_4$)

7.33 (m, 5 H, Ph)
4.68 (s, 1 H, −CHPh(OSiMe$_3$))
0.20 (s, 9 H, −OSiMe$_3$)

IR (cm^{-1}, CHCl$_3$)

M−C≡O stretches: 2119, 2048, 2020
C=O stretch: 1638

For **11**

^{1}H NMR (δ , CDCl$_3$)

7.2 (m, 5 H, Ph)
6.2 (s, 1 H, −CHPhOH)
5.0 (s, 1 H, −OH)

IR (cm^{-1}, CHCl$_3$)

O−H stretch: 3610−3260
M−C≡O stretches: 2116, 2052, 2016
C=O stretch: 1639

Selover, J.C.; Vaughn, G.D.; Strouse, C.E.; Gladysz, J.A. *J. Am. Chem. Soc.* **1986**, _108_, 1455.

3. A metallacyclobutane, **15**, has been invoked as an intermediate. β−hydrogen (or deuterium) abstraction from the metallacycle can give either the alkene-deuteride complex **14** or the deuterated alkene-hydride **16**. Reductive elimination from **14** and **16** gives the observed methyl vinyl ether derivatives.

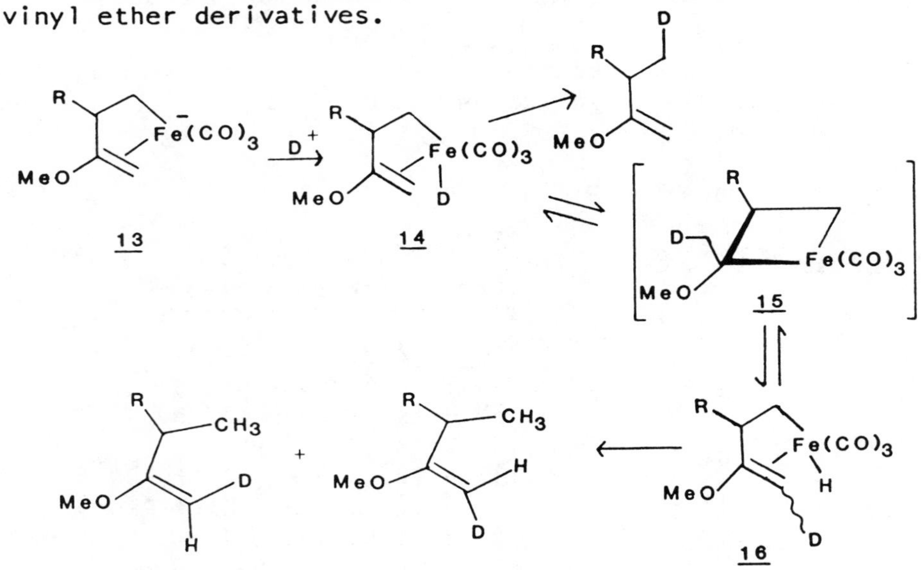

Semmelhack, M.F.; Le, H.T.M. *J. Am. Chem. Soc.* **1985**, _107_, 1455.

4. The authors believe that elimination of isobutylene (**18**) occurs from the metallacycle to give a nickel carbene complex as the reactive intermediate. Carbonylation of this carbene produces the ketene complex **17**, which can react with LiAlH$_4$ and methanol to give ethanol (**19**) and methyl acetate (**20**) respectively. In the scheme below, L = PPh$_3$.

Miyashita, A.; Shitara, H.; Nohira, H. _J_. _Chem_. _Soc_., _Chem_. _Comm_. **1985**, 850.

5. The proposed mechanism is shown on the following page. In all cases, L = PPh$_3$. Pi-coordination of two phenylallenes by the unsubstituted double bond occurs initially to give **34**. Stereospecific and regiospecific coupling then occurs to give metallacycle **35**. The stereochemistry and regiochemistry of this coupling is thought to be controlled by steric effects, i.e., the benzylidene rotates _away_ from the approaching PPh$_3$ ligand. Carbonylation of **35** gives the undetected intermediate **36** which isomerizes to give either **21** or **22**.

NOTE
Although mono- and _trans_ bis-pi-complexes must be present during this reaction, they do not lead to the products shown.

5. (cont.)

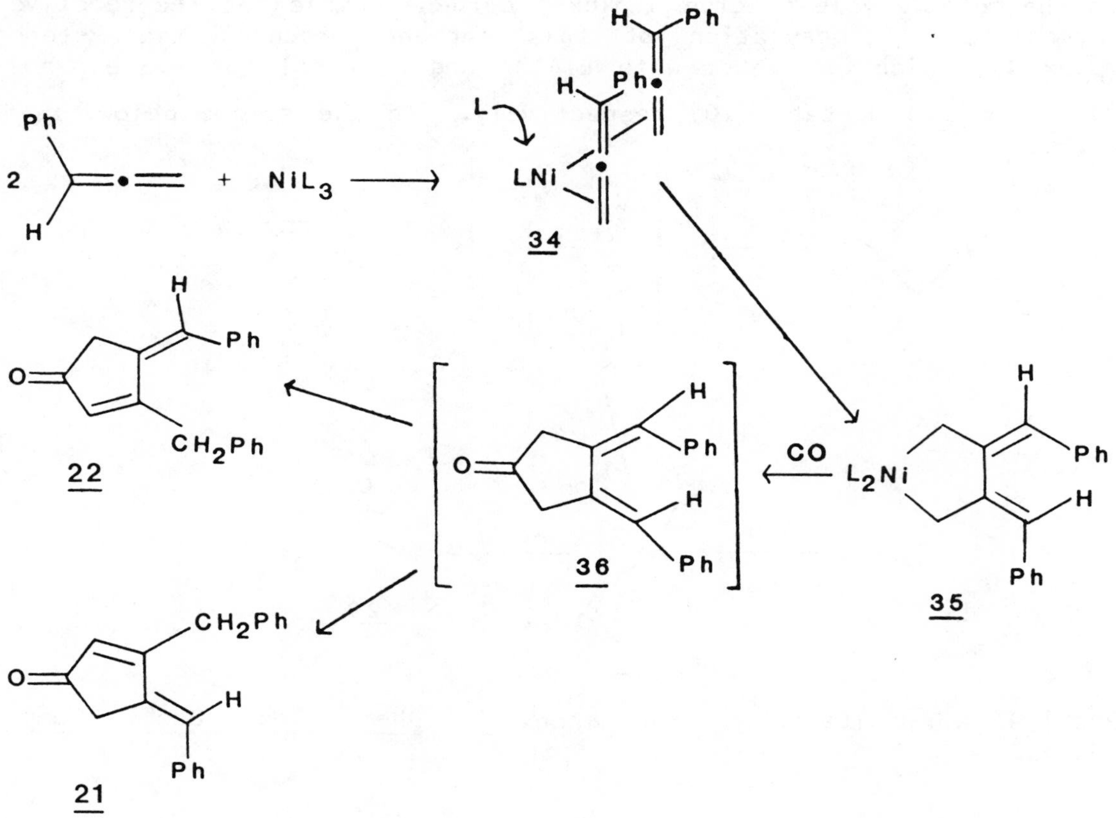

Pasto, D.J.; Huang, N.-Z.; Eigenbrot, C.W. J. Am. Chem. Soc. 1985, 107, 3160.

6. a) Complexes 23 and 24 are shown below:

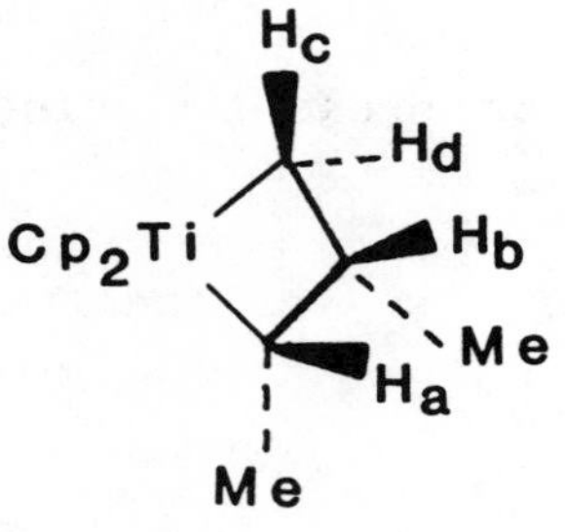

23

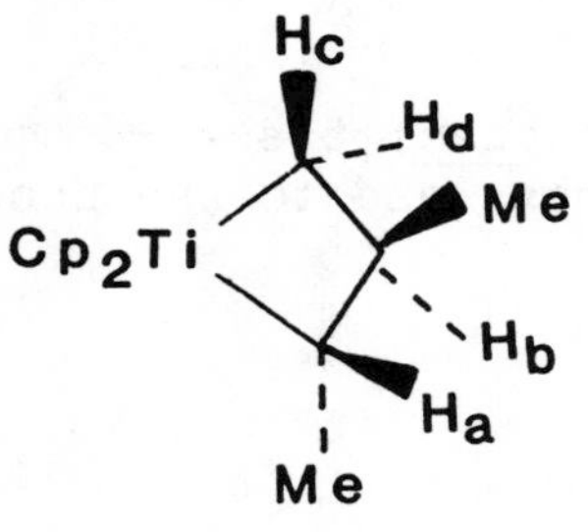

24

6.   a) (cont.) Assignments for ^{1}H NMR data (based on selective homonuclear decoupling and comparison to data for other β-substituted metallacycles):

For **23**

0.03 (m, 1 H, H$_b$)
1.02 (d, 3 H, -C$\underline{H}_3$)
1.58 (d, 3 H, -C$\underline{H}_3$)
2.31 (t, 1 H, H$_d$)
3.51 (m, 2 H, H$_a$, H$_c$)
5.23 (s, 5 H, Cp)
5.32 (s, 5 H, Cp)

For **24**

-0.80 (m, 1 H, H$_b$)
1.04 (d, 3 H, -C$\underline{H}_3$)
1.68 (d, 3 H, -C$\underline{H}_3$)
2.33 (dd, 1 H, H$_d$)
2.81 (dd, 1 H, H$_a$)
3.05 (dd, 1 H, H$_c$)
5.27 (s, 5 H, Cp)
5.36 (s, 5 H, Cp)

Note the retention of the stereochemistry of the starting olefin in the metallacycle. This appears to be a general result for all olefins tested.

b) The amine serves to drive the equilibrium to the right by removing $Al(CH_3)_2Cl$. The highly-reactive titanium methylene species then reacts rapidly with an olefin to give the metallacycle.

NOTES
1) The solubility of the $Me_2(Cl)Al-NR_3$ adduct (and hence the ease of separation of the metallacycle from the reaction mixture) is affected by the base used. Generally, the adduct precipitates from the reaction mixture to leave the metallacycle in solution; however, the DMAP-Al(Cl)Me$_2$ adduct and the cis-1,2-dimethyltitanacyclobutane, **23**, have similar solubility properties, and thus are difficult to separate and purify. If 4-t-butylpyridine is used as the base, the AlMe$_2$Cl-adduct remains soluble while complex **23** precipitates out.

6.   (cont.)
2) With weaker bases such as diethyl ether and THF, significant amounts of both the metallacyclic complexes and the starting material are present in solution.   Only with stronger bases such as DMAP and 4-t-butylpyridine does the reaction give only the metallacycle.

3)   An   interesting   aspect   of   the ^{1}H   NMR   spectra   of   the titanacyclobutanes is the unusual upfield shifts (0-1 ppm) observed for the β-protons.

Howard, T.R.; Lee, J.B.; Grubbs, R.H. J. Am. Chem. Soc. **1980**, 102, 6876.

Straus, D.A.; Grubbs, R.H. Organomet. **1982**, 1, 1658.

7.   The proposed mechanism is shown below.

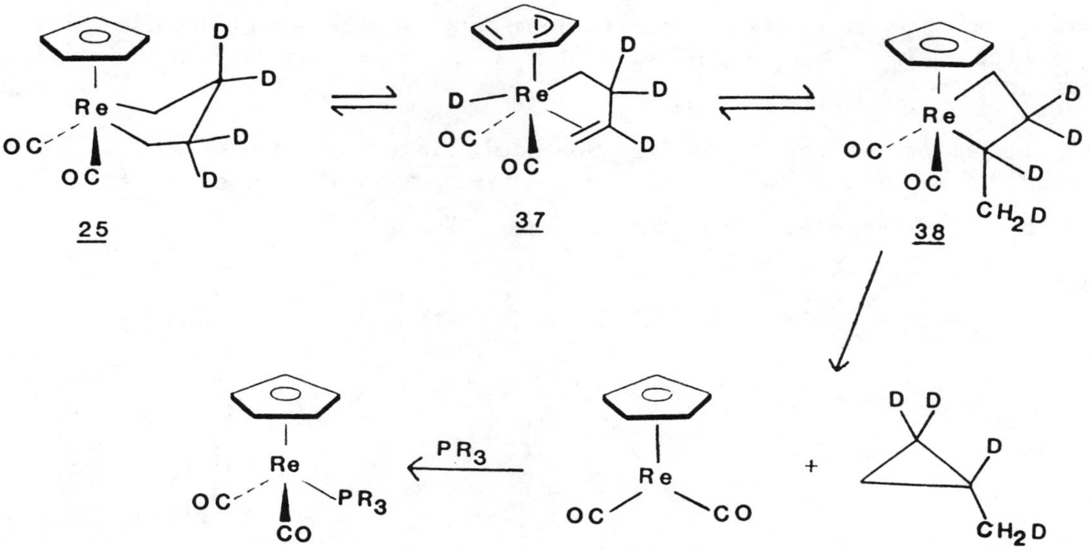

The first step is a β-H abstraction, with concomitant slippage of the Cp ring to η^3 to generate **37**.   This step is followed by olefin insertion to give the metallacyclobutane, **38**.   The last step is a reductive-elimination to give the cyclopropane.   An interesting aspect of this reaction is that the olefin-insertion/reductive-elimination is faster than reductive-elimination from **37** to give 1-butene.

Yang, G.K.; Bergman, R.G. Organomet. **1985**, 4, 129.

8.   In order to substantiate this mechanistic hypothesis, one could:

1) Obtain rate data and determine the molecularity of the reaction. The   possibility   that   reaction   of   **29**   with   styrene   is   the rate-determining step can be eliminated by this process.   If the first step is the rate-determining step, the data should fit a first-order plot, and the rate should not be affected by the addition of excess styrene.

8. (cont.)

2) Add labeled styrene, PhCH=CD$_2$, to a solution of **26** and look for exchange into the isomerized metallacycle, **27**, and into the starting complex.

3) Use a labeled titanacyclobutane, such as **39**, and look for the formation of **27**, **40**, **41**, and **42**.

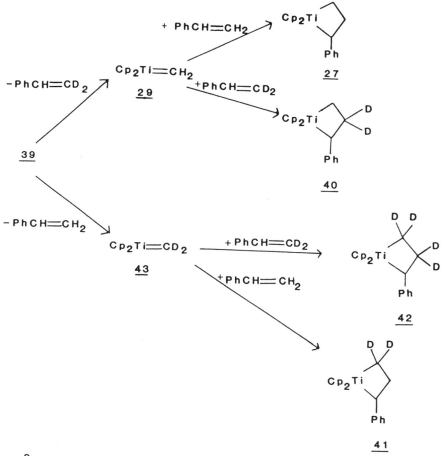

These products can be formed by the pathways shown below.

The use of ^{2}H NMR spectroscopy would greatly simplify the analysis of the products from this reaction. The signals are well separated and the spectrum will show only the deuterated products.

Ikariya, T.; Ho, S.C.H.; Grubbs, R.H. Organomet. **1985**, 4, 199.

9. The reaction to give complex **31** is shown below:

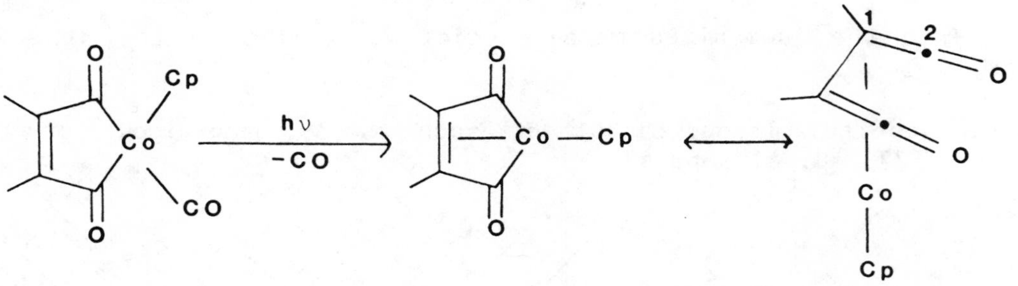

31

Assignments for spectral data for **31**:

^{1}H NMR (δ, ppm)

1.91 (s, 6 H, $-CH_3$)

5.02 (s, 5 H, Cp)

^{13}C NMR (δ, ppm)

10.8 ($-CH_3$)

51.3 (C_1)

86.8 (Cp)

225.9 (C_2)

Jewell, C.F.; Liebeskind, L.S.; Williamson, M. J. Am. Chem. Soc. **1985**, 107, 6715.

10. The proposed mechanism involves initial dissociation of PEt_3, followed by oxidative-addition of a C–H bond of one of the neopentyl ligands. Reductive-elimination of neopentane and reassociation of PEt_3 gives metallacycle **33**.

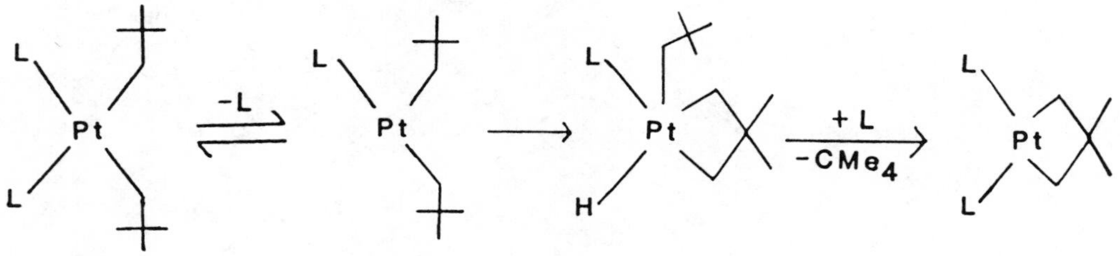

32   **33**

Foley, P.; Whitesides, G.M. J. Am. Chem. Soc. **1979**, 101, 273.

10

Reactions of Transition Metal Alkyl and Hydride Complexes

$$Mn_2(CO)_9 \quad + \quad CO \quad \xrightarrow{\text{fast}} \quad Mn_2(CO)_{10}$$

QUESTIONS

1. Predict the organic products of the following reactions:

a) $Fe(CO)_5$ + KOH + [methyl vinyl ketone] $\xrightarrow{\text{EtOH}}$

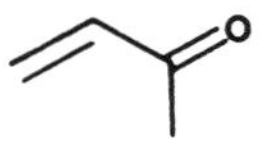

b) $[HFe(CO)_4]^-$ + [methyl vinyl ketone] $\xrightarrow{\text{THF}}$

c) $[HFe(CO)_4]^-$ + [benzoyl chloride] $\xrightarrow{CH_2Cl_2}$

d) $[HFe_2(CO)_8]^-$ + [substituted cyclohexenone] $\xrightarrow{\text{HOAc}}$

e) $[HW(CO)_5]^-$ + [bromonorbornane] $\xrightarrow{\hspace{1cm}}$

f) $[HCr(CO)_5]^-$ + [Br–(CH$_2$)$_3$–C(O)Cl] $\xrightarrow{\hspace{1cm}}$

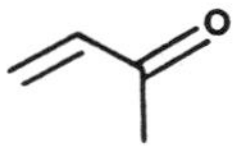

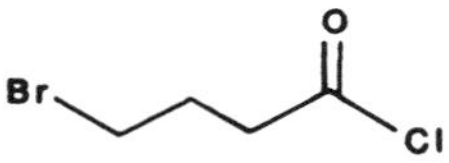

g) $(PPh_3)_2CuBH_4$ + [furan-2-carbonyl chloride] $\xrightarrow{\hspace{1cm}}$

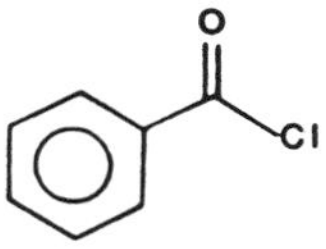

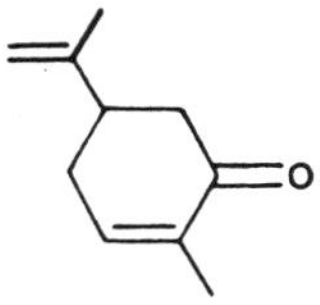

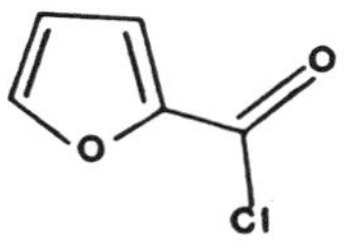

2. Suggest an organometallic reagent to carry out the following transformations.

a) [PhCH=CH–C(O)CH$_3$] $\longrightarrow$ [PhCH$_2$CH$_2$–C(O)CH$_3$]

b) [CH$_3$(CH$_2$)$_3$C(O)Cl] $\longrightarrow$ [CH$_3$(CH$_2$)$_3$C(O)H]

c) [3-bromocyclohexanone] $\longrightarrow$ [cyclohexanone]

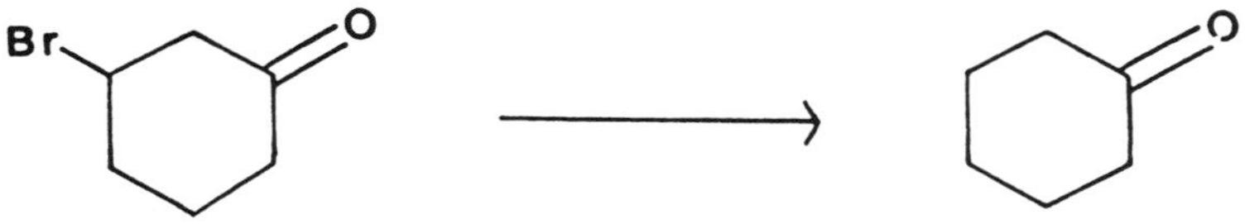

2. d)

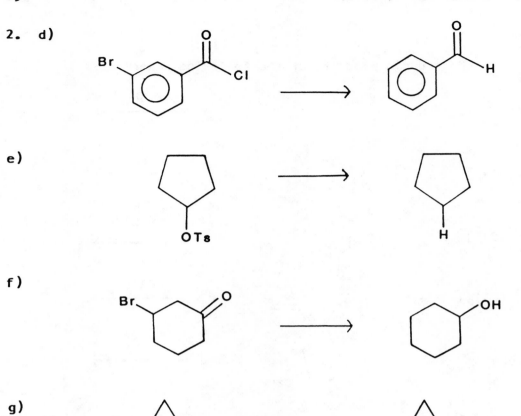

e)

f)

g)

h)

3. Predict the products of the reaction (if any) of $[CpV(CO)_3H]^-$ with the following substrates.

a)

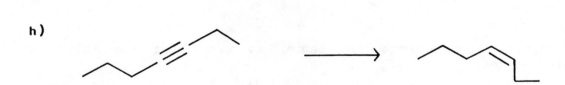

b)

3. **c)**

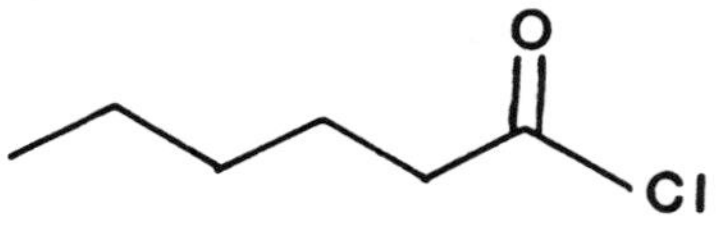

d)

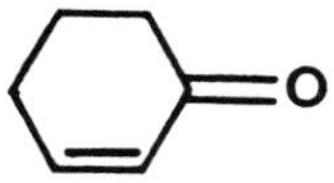

e)

f)

g)

4. List six ways to make metal-carbon sigma bonds (excluding M-CO bonds) and give an example of each.

5. Predict the products of the following reactions:

a) $Cp_2Zr(H)Cl$ +

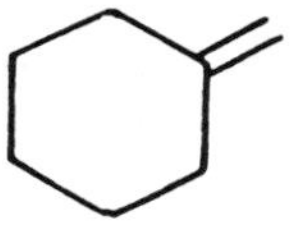

b) $Cp_2Zr(H)Cl$ +

c)

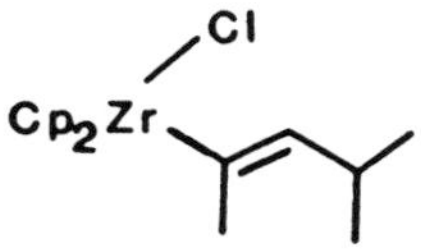

+ N-bromosuccinimide $\longrightarrow$ organic product

d)

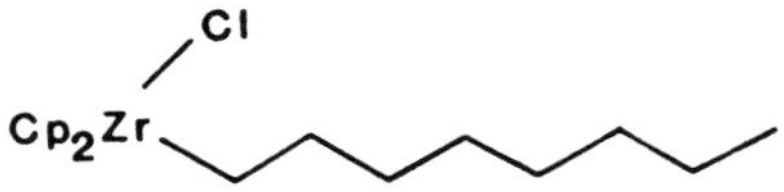

+ t-BuOOH $\longrightarrow$ organic product

e)

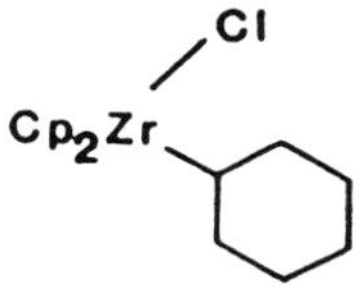

+ CO $\longrightarrow$

f)

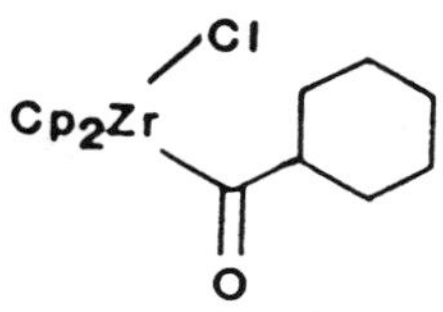

+ H_2O_2 $\longrightarrow$ organic product

6. Derive the rate expression for the formation of **5** in terms of measurable concentrations for the following reaction.

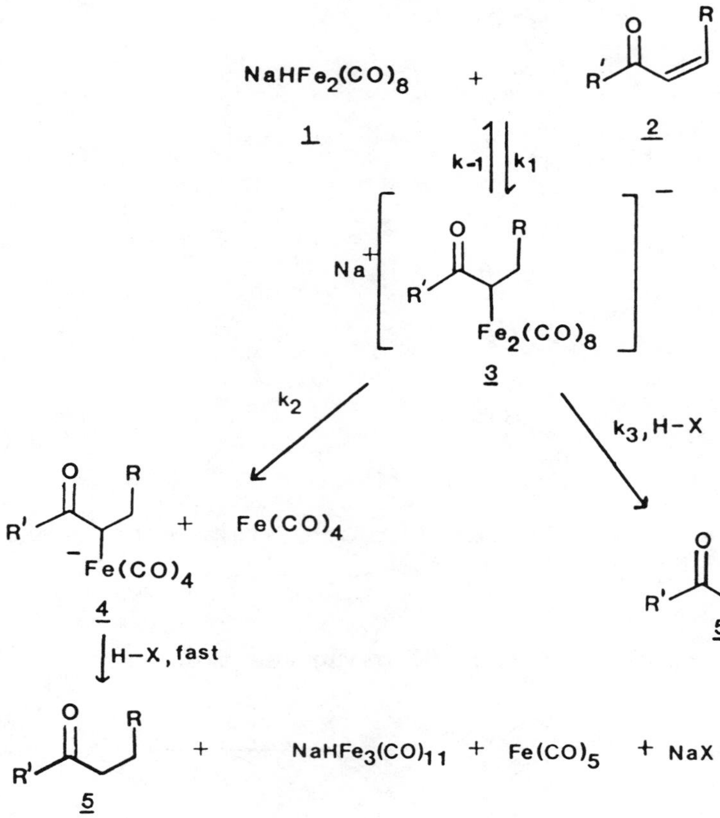

$$NaHFe_2(CO)_8 \quad + \quad$$

1

$$k_{-1} \Big\Uparrow k_1$$

2

$$Na^+ \left[\begin{array}{c} \text{[structure 3]} \\ Fe_2(CO)_8 \end{array} \right]^-$$

3

$$k_2 \swarrow \qquad \qquad \searrow k_3, H-X$$

$+ \quad Fe(CO)_4$

4

$$\Big\downarrow H-X, fast$$

$+ \quad NaHFe_3(CO)_{11} \quad + \quad Fe(CO)_5 \quad + \quad NaX$

5

5

7. The following rearrangement has been observed (L is a phosphine ligand and PF_6^- is the counterion in all cases):

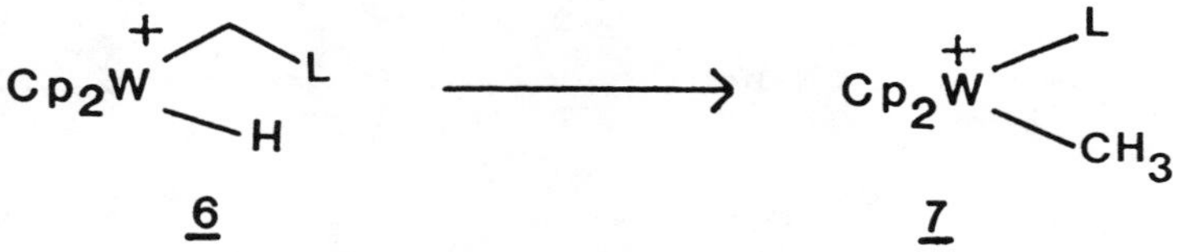

$\longrightarrow$

6   **7**

OBSERVATIONS

1) The rearrangement of **6** to **7** was found to follow first-order kinetics, i.e., $-d[6]/dt = k_{obs}[6]$.

2) $\Delta H^{\ddagger} = 141.7$ kJ/mol and $\Delta S^{\ddagger} = 61.7$ J/mol K.

3) The starting material, **6**, readily exchanges phosphine ligand L for a different phosphine, L'.

4) Reaction of equimolar amounts of **6**-d_3 and **6**-d_0 generates only **7**-d_3 and **7**-d_0.

5) The rate of the reaction is independent of added phosphine concentration.

7. (cont.) The authors propose the two different mechanisms, outlined below:

Mechanism I

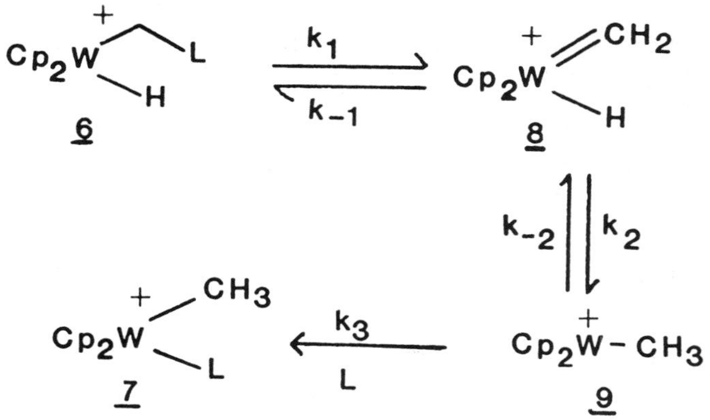

Mechanism II

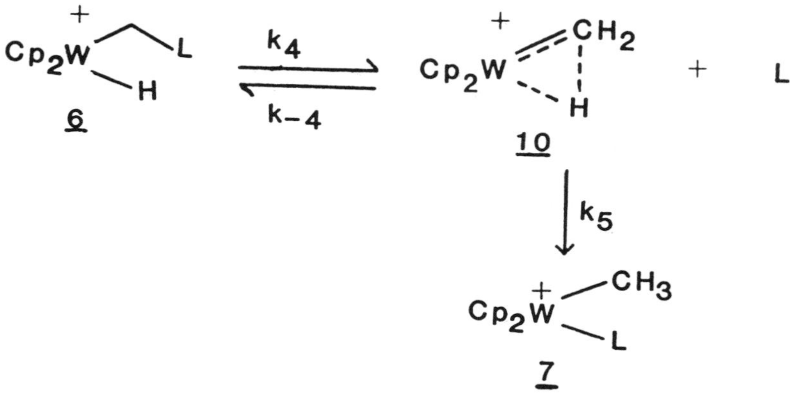

a) For each mechanism, write a kinetic expression for the disappearance of **6** in terms of the rate constants given. Assume all intermediates are at a steady-state concentration.

b) Can these two mechanisms be kinetically distinguished by the fact that there is no inverse dependence of the rate on the phosphine concentration? Why or why not?

8. The following reaction occurs:

$$2 \; Cp(CO)_3Mo\text{--}H \;+\; PhC(CH_3)\text{=}CH_2 \longrightarrow [Cp(CO)_3Mo]_2 \;+\; PhCH(CH_3)_2$$

$$\underline{11} \qquad\qquad\qquad \underline{12}$$

OBSERVATIONS
1) Second order kinetics are observed, i.e.,

$$-d[\mathbf{12}]/dt = k_{obs}[\mathbf{11}][\mathbf{12}]$$

8. (cont.)

2) The rate is not affected by 1 atmosphere of CO.

3) An <u>inverse</u> isotope effect is observed, i.e. $Cp(CO)_3M-D$ reacts faster than $Cp(CO)_3M-H$. No D-incorporation into the methyl group of unreacted **12** is observed.

Postulate a mechanism consistent with all observations.

9. Below is a series of hydride reducing agents listed in order of decreasing reactivity towards primary alkyl halides:

$$[M-H]^- + R-X \longrightarrow [M-X]^- + R-H$$

$[HW(CO)_4(P(OMe)_3)]^- > [HCr(CO)_4(P(OMe)_3)]^- > [HW(CO)_5]^- > [CpV(CO)_3H]^-$

$> [HCr(CO)_5]^- > [HRu(CO)_4]^- > [\underline{trans}-HFe(CO)_3(P(OMe)_3)]^- \gg [HFe(CO)_4]^-$

Two mechanistic pathways are possible:
1) Ionic (S_N2-like) hydride transfer, or

2) single electron transfer, followed by hydrogen atom abstraction.

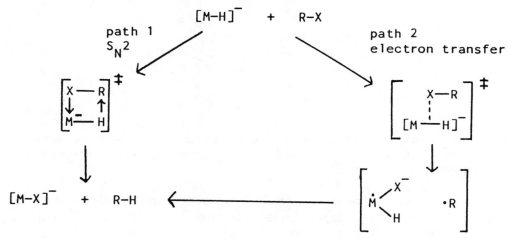

Based on the following observations, decide which pathway is favored for each of the two complexes, $[HW(CO)_4P]^-$ (P=P(OMe)$_3$) and $[HW(CO)_5]^-$. Explain all observations in light of your conclusions.

OBSERVATIONS

1) Reaction of $[HW(CO)_4P]^-$ and $[HW(CO)_5]^-$ with 6-bromo-1-hexene gives the mixtures of products shown.

9. (cont.)

[M-H]⁻   +   ⟶   +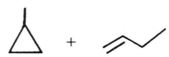

[M-H]⁻	Product Ratio, 1-hexene:methylcyclopentane
[HW(CO)₄P]⁻	1 : 0
[HW(CO)₅]⁻	1 : 0.67

2) Reaction of [HW(CO)₄P]⁻ and [HW(CO)₅]⁻ with cyclopropylcarbinyl bromide gives the mixtures of products shown.

[M-H]⁻   +   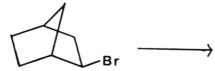   ⟶   +

[M-H]⁻	Product Ratio, methylcyclopropane:1-butene
[HW(CO)₄P]⁻	1 : 0.1
[HW(CO)₅]⁻	1 : 3.5
(n-Bu)₃Sn-H	1 : 200

3) The half-lives for displacement of tosylate from n-butyltosylate are:

[M-H]⁻	$t_{1/2}$
[HW(CO)₄P]⁻	22 min.
[HW(CO)₅]⁻	> 8 h
[HCr(CO)₅]⁻	33 h

4) Reaction of [DW(CO)₄P]⁻ and [DW(CO)₅]⁻ with <u>exo</u>-2-bromonorbornane gives the products shown.

M-D⁻   +      ⟶   +

[M-H]⁻	Product ratio, exo/endo-2-deuterionorbornane
[DW(CO)₄P]⁻	1 : 0.43
[DW(CO)₅]⁻	1 : 0
[Et₃BD]⁻	0 : 1

10. There are a number of possible reaction mechanisms that have been postulated for binuclear reductive-elimination reactions between transition metal alkyl and transition metal hydride complexes.

$$M-R \quad + \quad M'-H \quad \longrightarrow \quad R-H \text{ or } RCHO$$

These mechanisms are:
1) Migration of the alkyl moiety to coordinated CO, followed by reductive-elimination of the alkane;

2) migration of the alkyl moiety to coordinated CO, followed by reductive-elimination of the aldehyde;

3) dissociation of CO, followed by elimination of the alkane;

4) homolytic cleavage of the M-R bond, followed by H-atom abstraction;

5) insertion of CO into the metal-alkyl bond, followed by homolytic cleavage of the metal-acyl bond and H-atom abstraction.

Reactions of $HMn(CO)_5$ with manganese alkyls have been stuc_d in order to determine the mechanism(s) for binuclear reductive-elimination in these systems. From the data given for the following two reactions, classify each one into one of the given categories.

10. (cont.)

Reaction 1

$$(CO)_5Mn-CH_2C_6H_4-p-OMe \quad + \quad (CO)_5Mn-H$$

13   **14**

$$\downarrow \quad CH_3CN, \ 20-30^{\circ}C$$

$$p-MeOC_6H_4CH_2CHO \quad + \quad Mn_2(CO)_9(NCCH_3)$$

15   **16**

OBSERVATIONS

1) Linear plots of $1/k_{obs}$ vs $1/[HMn(CO)_5]$ are obtained.

2) Reaction of **13** with PMe_2Ph results in the formation of $(CO)_4(PMe_2Ph)Mn-COCH_2C_6H_4-p-OMe$.

3) The activation parameters for the reaction are: $\Delta H^{\ddagger} = 17.8$ kcal/mol; $\Delta S^{\ddagger} = -14$ eu.

4) Reaction of **13** with **14** in benzene gives no **15** or **16**, but instead gives the alkane, $p-MeOC_6H_4CH_3$, and $Mn_2(CO)_{10}$.

Reaction 2

$$\underline{cis}-p-MeOC_6H_4CH_2-Mn(CO)_4P \quad + \quad HMn(CO)_5$$

17   **14**

$P = (p-MeO-C_6H_4)_3P$ $\downarrow$ benzene, $75^{\circ}C$

$$Mn_2(CO)_{10} \quad + \quad HMn(CO)_4P \quad + \quad Mn_2(CO)_9P \quad + \quad p-MeOC_6H_4CH_3$$

18   **19**   **20**   **21**

FACTS

1) The pseudo-first order rate constant k_{obs} is independent of the concentration of $HMn(CO)_5$ (**14**).

2) The amount of $HMn(CO)_4P$ (**19**) obtained is very dependent on the concentration of $HMn(CO)_5$ (**14**), as shown in the table on the following page. Compound **19** is not formed by direct reaction of $Mn_2(CO)_9P$ (**20**) with **14**.

10. (cont.)

[17], M	[14], M	% 19
0.026	0.30	77
0.023	0.60	87
0.025	0.90	90
0.025	1.20	94

3) Reaction of **17** with $Mn_2(CO)_8(PMePh_2)_2$ gives a mixture of $p-MeOC_6H_4CH_2-Mn(CO)_4(PMePh_2)$ and $Mn_2(CO)_8(PMePh_2)[P(C_6H_4-p-OMe)_3]$

4) Complex **17** does not lose CO when heated to 75°C.

11. The following reaction sequence can be used to obtain diallyl ketones.

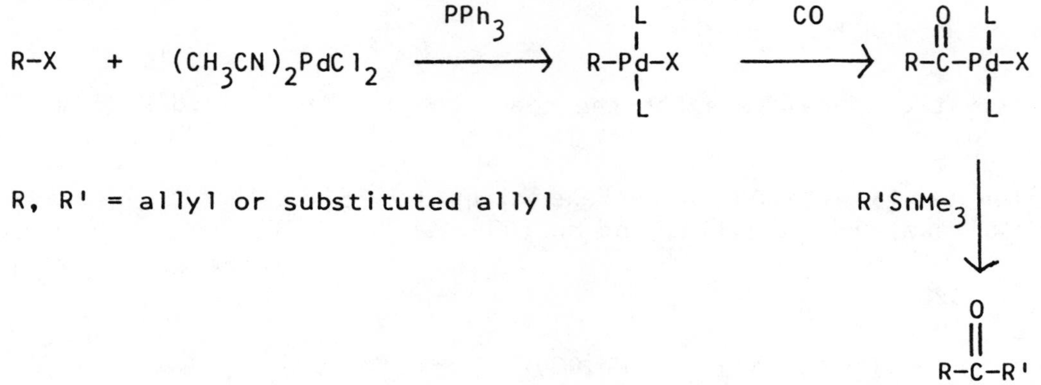

$$R-X \ + \ (CH_3CN)_2PdCl_2 \ \xrightarrow{PPh_3} \ R-\underset{\underset{L}{|}}{\overset{\overset{L}{|}}{Pd}}-X \ \xrightarrow{CO} \ R-\underset{\underset{L}{|}}{\overset{\overset{O \quad L}{|| \quad |}}{C-Pd}}-X$$

R, R' = allyl or substituted allyl

$$\Bigg\downarrow R'SnMe_3$$

$$\underset{R-C-R'}{\overset{O}{\overset{||}{}}}$$

Using this transmetallation reaction, design syntheses for the following compounds.

a)

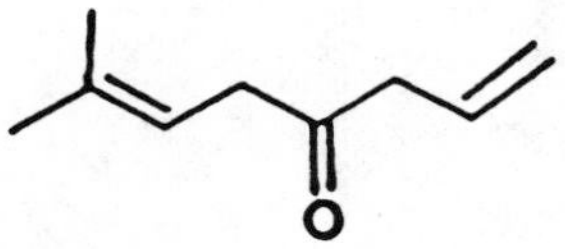

b)

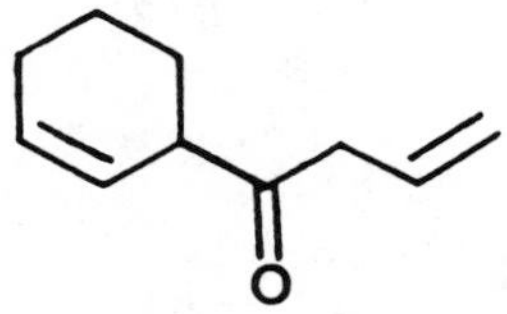

c)

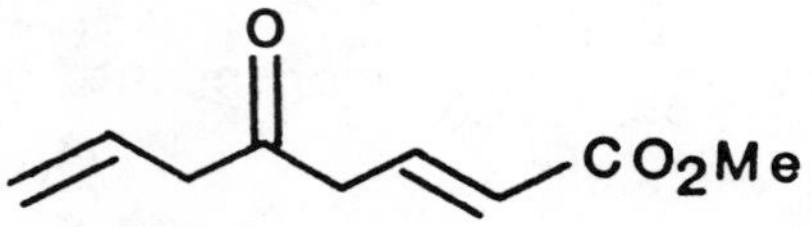

ANSWERS

1. The organic products are shown below:

a)

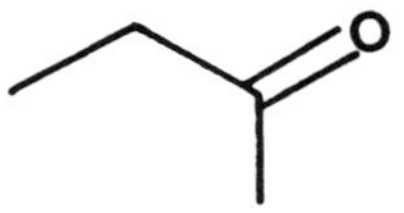

b)

c)

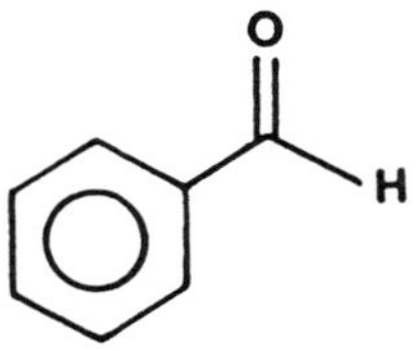

d)

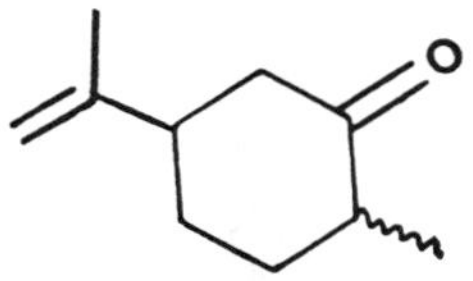

e)

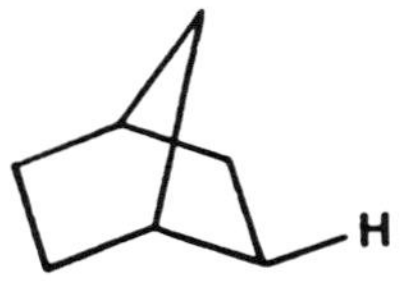

f)

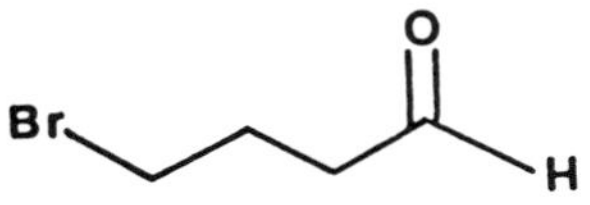

g)

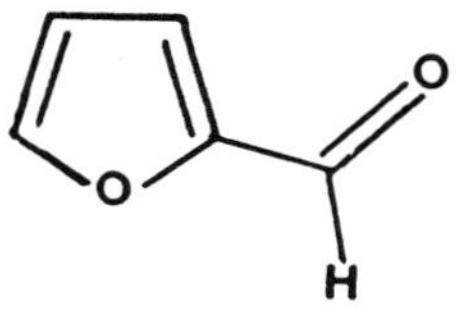

Collman, J.P.; Hegedus, L.S.; Norton, J.R.; Finke, R.G. "Principles and Applications of Organotransition Metal Chemistry", 2nd ed., University Science Books (Mill Valley, CA), Chapter 13.

2. The reagents given are only those discussed in Chapter 13 of Collman, Hegedus, Norton, and Finke's "Principles and Applications of Organotransition Metal Chemistry", 2nd edition, University Science Books (Mill Valley, CA). There may also be other reagents that effect the same transformations.

a) Possible reagents are $[HFe(CO)_4]^-$ in a protic solvent, $[HFe_2(CO)_8]^-$, $[HCr_2(CO)_{10}]^-$, $[n\text{-}Pr\text{-}C\equiv C\text{-}Cu\text{-}H]^-$, and $Bu_3Sn\text{-}H/Pd(0)$.

b) Possible reagents are $[HFe(CO)_4]^-$ in an aprotic solvent, $[HCr(CO)_5]^-$, $[CpV(CO)_3H]^-$, and $Bu_3Sn\text{-}H/Pd(0)$.

c) Possible reagents are $[HCr(CO)_5]^-$, $[HW(CO)_5]^-$, and $[CpV(CO)_3H]^-$.

d) Possible reagents are $[HCr(CO)_5]^-$ and $[HW(CO)_5]^-$.

e) Possible reagents are $[n\text{-}BuCuH]^-$ and $LiAlH_4/NiCl_2$.

2. f) [n–BuCuH]⁻.

g) [CpV(CO)₃H]⁻.

h) LiAlH₄/NiCl₂.

3. a)

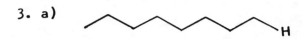

b) no reaction

c)

d) no reaction

e) no reaction   **f)**

g) no reaction

Kinney, R.J.; Jones, W.D.; Bergman, R.G. <u>J</u>. <u>Am</u>. <u>Chem</u>. <u>Soc</u>. **1978**, 100, 7902.

4. Six possible ways to form metal–carbon sigma bonds are given below. A great number of examples of these types of reactions exist, one of which is given for each reaction type.

1) M⁻   +   R–X (X = leaving group) ⟶ M–R

Example: Fp⁻   +   CH₃–I ⟶ Fp–CH₃

King, R.B. "Organometallic Syntheses", vol. 1, Academic Press (New York, 1965), p. 151.

2) M–X (X = halide)   +   R–MgX or R–Li ⟶ M–R

Example: (MeCp)₂ZrCl₂   +   2 MeLi ⟶ (MeCp)₂ZrMe₂

Martin, B.D.; Matchett, S.A.; Norton, J.R.; Anderson, O.P. <u>J</u>. <u>Am</u>. <u>Chem</u>. <u>Soc</u>. **1985**, 107, 7952.

3) L$_n$M   +   R–X (X = halide or tosylate) ⟶ L$_n$M(R)(X)

This type of oxidative-addition reaction occurs most frequently with low-valent, electron-rich metals such as Ni(0), Pd(0), Ir(I), etc.

Example:
Pd(OAc)₂   +   2 PPh₃   +   p-NO₂–C₆H₄Br ⟶ (PPh₃)₂Pd(Br)(C₆H₄–p-NO₂)

Heck, R.F. <u>Pure</u> <u>Appl</u>. <u>Chem</u>. **1978**, 50, 691.

4. (cont.)

4)

M—H + CH₂=CH—R ⟶ M—CH₂CH₂—R (with structures drawn)

Example:

Cp₂Nb(=CH₂)(H) ⇌ Cp₂Nb(ethyl)

Note that this reaction is an equilibrium. The niobium–ethyl complex is not isolable.

Doherty, N.M.; Bercaw, J.E. J. Am. Chem. Soc. **1985**, 107, 2670.

5) M–H + CH₂=CH–R ⟶ M–CH₂CH₂CH₂–R

This type of reaction only occurs with electron-poor, early transition metals.

Example: Cp₂ZrH(Cl)   +   1-pentene ⟶ Cp₂Zr(Cl)(n-pentyl)

Schwartz, J.; Labinger, J.A. Angew. Chem. Int. Ed. Engl. **1976**, 15, 333.

6) M—‖(R)   +   H⁻ ⟶ M

Example:

[CpMo(CO)₃(ethylene)]⁺ + NaBH₄ ⟶ CpMo(CO)₃(ethyl)

Cousins, M.; Green, M.L.H. J. Chem. Soc. **1963**, 889.

5. a)

Cp₂Zr(Cl)(CH₂-cyclohexyl)

b)

Cp₂Zr(Cl)(n-decyl)

c)

Br–C(CH₃)=CH–CH(CH₃)₂

d)

HO-(octyl chain)

e)

Cp₂Zr(Cl)(C(=O)-cyclohexyl)

f)

cyclohexyl-C(=O)-OH

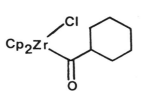

Schwartz, J.; Labinger, J.A. Angew. Chem. Int. Ed. Engl. **1976**, 15, 333.

6. The rate expression for the formation of **5** is:

$$d[5]/dt = k_2[3] + k_3[3][H-X] = [3](k_2 + k_3[H-X])$$

Since [3] is not a measurable quantity, we must use the equilibrium between **1** and **3** to find an expression for [3].

$$\frac{k_1}{k_{-1}} = K_{eq} = \frac{[3]}{[1][2]}$$

and thus [3] = K_{eq}[1][2].

Substituting this value for [3] into the initial rate expression, we obtain a rate expression in which all concentrations are measurable,

$$d[5]/dt = K_{eq}[1][2](k_2 + k_3[H-X])$$

Collman, J.P.; Finke, R.G.; Matlock, P.L.; Wahren, R.; Komoto, R.G.; Brauman, J.I. J. Am. Chem. Soc. **1978**, 100, 1119.

7. a) For Mechanism I, the derivation of the rate expression requires the use of two steady-state approximations.

$$6 \quad \underset{k_{-1}}{\overset{k_1}{\rightleftharpoons}} \quad 8 + L$$

$$8 \quad \underset{k_{-2}}{\overset{k_2}{\rightleftharpoons}} \quad 9$$

$$9 + L \quad \overset{k_3}{\longrightarrow} \quad 7$$

Overall, the rate is given by equation (1)

$$-d[6]/dt = k_3[9][L] = d[7]/dt \tag{1}$$

Using the steady-state approximation for **9**, we find that

$$d[9]/dt = k_2[8] - k_3[9][L] - k_{-2}[9] = 0 \tag{2}$$

and $\quad [9] = \dfrac{k_2[8]}{k_{-2} + k_3[L]} \tag{3}$

This expression still contains an unmeasurable quantity, [8]; however, since **8** is also a steady-state intermediate, we can use the steady-state approximation to obtain an expression for [8].

$$d[8]/dt = k_1[6] + k_{-2}[9] - k_{-1}[L][8] - k_2[8] = 0 \tag{4}$$

7. (cont.)

and $\quad [8] = \dfrac{k_1[6] + k_{-2}[9]}{k_{-1}[L] + k_2}$ $\qquad\qquad\qquad\qquad$ (5)

Combination of these two equations (a somewhat involved, but straightforward algebraic manipulation) will give an expression for **9** that can be substituted into equation (1) to give the rate expression

$$\frac{-d[6]}{dt} = \frac{k_1 k_2 k_3 [6]}{k_{-1}k_{-2} + k_2 k_3 + k_{-1}k_3[L]} \qquad\qquad (6)$$

For Mechanism II, there is only one steady-state intermediate, **10**, which makes the kinetic analysis simpler.

$$6 \quad \underset{k_{-4}}{\overset{k_4}{\rightleftharpoons}} \quad 10 + L$$

$$10 + L \quad \xrightarrow{k_5} \quad 7$$

The rate of the reaction is given by equation (7)

$$-d[6]/dt = k_5[10][L] = d[7]/dt \qquad\qquad (7)$$

A steady-state approximation for **10** gives:

$$d[10]/dt = k_4[6] - k_{-4}[10][L] - k_5[10][L] \qquad\qquad (8)$$

and $\quad [10] = \dfrac{k_4[6]}{[L](k_{-4} + k_5)}$ $\qquad\qquad\qquad\qquad$ (9)

Substituting this expression for [10] back into equation (7) gives the rate expression

$$\frac{-d[6]}{dt} = \frac{k_4 k_5 [6]}{k_{-4} + k_5} \qquad\qquad (10)$$

b) Initially it might be thought that a lack of an inverse dependence on L rules out Mechanism I; however, if the equilibrium between **8** and **9** is established very quickly (i.e., k_2 and k_{-2} are large), then $k_{-1}k_{-2} + k_2 k_3 \gg k_{-1}k_3[L]$, and the rate expression for Mechanism I reduces to

$$\frac{-d[6]}{dt} = \frac{k_1 k_2 k_3 [6]}{k_{-1}k_{-2} + k_2 k_3}$$

As a result, the fact that the rate does not depend on phosphine concentration does not enable one to distinguish between these two mechanisms.

Green, J.C.; Green, M.L.H.; Morley, C.P. _Organomet._ 1985, _4_, 1302.

8. The postulated mechanism is shown below:

$$Cp(CO)_3Mo-H \quad + \quad Ph(CH_3)C=CH_2 \quad \underset{k_2}{\overset{k_1}{\rightleftharpoons}} \quad [Cp(CO)_3Mo]\cdot \quad + \quad Ph(CH_3)_2C\cdot$$

$$\underline{11} \qquad\qquad\qquad \underline{12}$$

$$Cp(CO)_3Mo-H \quad + \quad Ph(CH_3)_2C\cdot \quad \underset{k_4}{\overset{k_3}{\rightleftharpoons}} \quad [Cp(CO)_3Mo]\cdot \quad + \quad Ph(CH_3)_2C-H$$

$$[Cp(CO)_3Mo]\cdot \quad + \quad [Cp(CO)_3Mo]\cdot \quad \underset{k_6}{\overset{k_5}{\rightleftharpoons}} \quad [Cp(CO)_3Mo]_2$$

The rate constants k_4 and k_6 must be negligible because of the endothermicity of these processes. The absence of deuterium exchange between $Cp(CO)_3Mo-D$ and the olefin, **12**, implies that the k_2 step is slow relative to the k_3 step. Thus, the rate-determining step must be the k_1 step.

The inverse isotope effect can be explained as follows: The zero point energy of the M—H bond is greater than that of the C—H bond, and therefore the transition state is unsymmetrical, as shown in **A**, below. When D is substituted for H, the zero point energies of both M—D and C—D are less than the corresponding M—H and C—H bond energies; however, because of the lighter mass of C relative to M, the difference is greater for C—D than for M—D. Thus the transition state becomes more unsymmetrical and is further along the reaction coordinate for M—D bond-breaking, as shown in **B**. Thus, the zero point energy difference between the transition states is <u>greater</u> than between the reactants, producing an inverse isotope effect.

$$M-----H---C \qquad\qquad\qquad M------D--C$$

$$\textbf{A} \qquad\qquad\qquad\qquad\qquad \textbf{B}$$

Sweany, R.L.; Comberrel, D.S.; Dombourian, M.F.; Peters, N.A. <u>J. Organomet. Chem.</u> **1981**, <u>216</u>, 57.

9. The phosphite-substituted hydride $HW(CO)_4[P(OMe)_3]^-$ appears to reduce alkyl halides by more of an ionic mechanism, while reductions using $[HW(CO)_5]^-$ have more radical character. The observations support this hypothesis:

1) The fact that methylcyclopentane is observed as a product in the

9. (cont.) reaction of $[HW(CO)_5]^-$ with 6-bromo-1-hexene implies that the following reaction is occurring:

$$k = 1 \times 10^5 s^{-1}$$

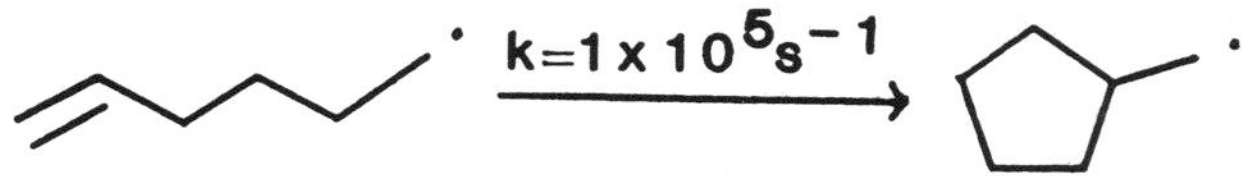

This means that alkyl radicals must be formed in the reaction of $[HW(CO)_5]^-$ with this substrate. The fact that no methylcyclopentane was observed for the reaction of $[HW(CO)_4(P(OMe)_3)]^-$ with this alkyl halide implies that ionic S_N2-type hydride transfer is occurring or that under these reaction conditions, the radical has a lifetime of less than 10^{-5} seconds.

2) The observation of ring-opened products from the reaction of $[HW(CO)_5]^-$ with cyclopropylcarbinyl bromide implies that the following reaction is occurring:

$$k = 1 \times 10^8 s^{-1}$$

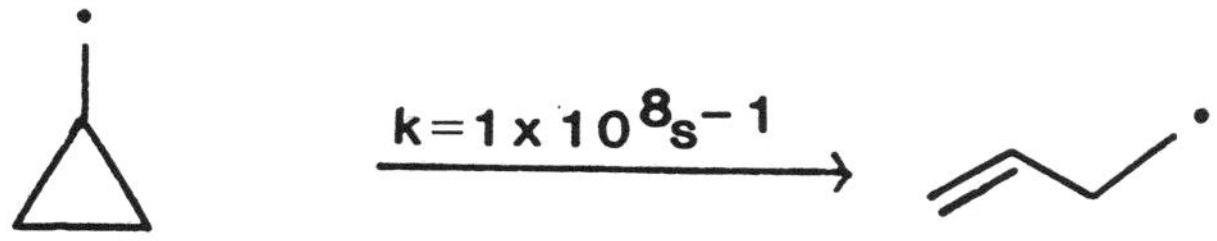

Bu_3Sn-H is known to react via this radical pathway. Since very little ring-opened product is observed for the reaction of $HW(CO)_4[P(OMe)_3]^-$ with cyclopropylcarbinyl bromide, the S_N2-type mechanism again seems to prevail for this species or under the reaction conditions, the radical lifetime is actually less than 10^{-8} seconds.

3) $LiBEt_3D$ is known to reduce alkyl halides via an ionic mechanism, which results in the inversion of the stereochemistry at the substituted carbon. The fact that $[DW(CO)_5]^-$ reacts with 2-bromonorbornane to give only the product with complete retention of configuration means that deuterium was delivered selectively to the less hindered _exo_ side of the molecule. This result is consistent with a radical mechanism. The fact that $[HW(CO)_4(P(OMe)_3)]^-$ gives 30% of the product with deuterium in the _endo_ position implies that some of the reaction could be going by an S_N2-type pathway with this reducing agent, and some by a radical mechanism. This might be anticipated, since this is a secondary alkyl bromide which does not undergo S_N2 processes readily.

Kao, S.C.; Spillett, C.T.; Ash, C.; Lusk, R.; Park, Y.K.; Darensbourg, M.Y. Organomet. **1985**, _4_, 83.

10. Reaction 1

The possible reaction pathways that yield aldehydes as products are 2) and 5).

2)

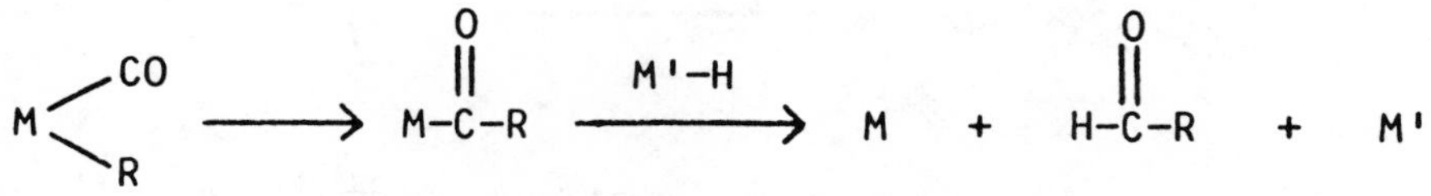

5)

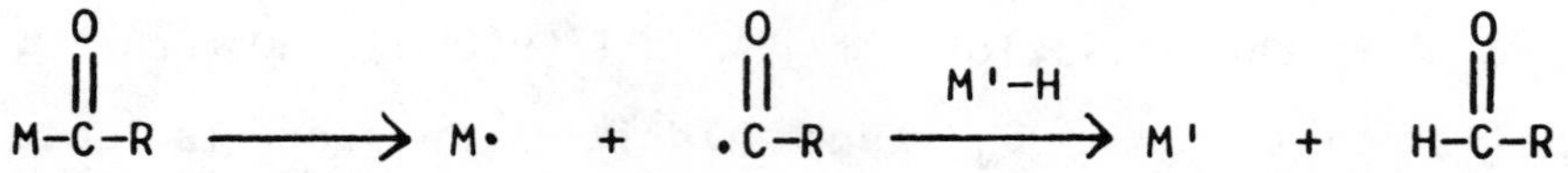

Because $\Delta S^{\ddagger}$ is negative, Mechanism 5 seems unlikely, since it involves a dissociative rate-determining step. The observation of completely different products in benzene indicates that the donor nature of the solvent is quite important, probably in the stabilization of a coordinatively unsaturated intermediate. The postulated mechanism, involving solvent-assisted CO-insertion, is shown below (R = $-CH_2C_6H_4-p-OMe$; S = CH_3CN):

$$(CO)_5Mn-R \quad + \quad S \quad \underset{k_{-1}}{\overset{k_1}{\rightleftharpoons}} \quad (CO)_4(S)Mn-COR$$

$$\underline{13} \hspace{10cm} \underline{21}$$

$$\underline{21} \quad + \quad HMn(CO)_5 \quad \xrightarrow{k_2} \quad RCHO \quad + \quad Mn_2(CO)_9(S)$$

$$\underline{14} \hspace{4cm} \underline{15} \hspace{3cm} \underline{16}$$

Using a steady-state approximation, the rate law shown in equation (11) can be derived.

$$\frac{-d[13]}{dt} = \frac{k_1 k_2 [13][14][S]}{k_{-1} + k_2[14]} \tag{11}$$

Since the solvent concentration is large, and therefore essentially constant, equation (11) simplifies to equation (12).

$$\frac{-d[13]}{dt} = \frac{k_1 k_2 [13][14]}{k_{-1} + k_2[14]} \tag{12}$$

This equation can be simplified further by dividing both sides by [13] to give

$$\frac{-d[13]}{[13]dt} = \frac{k_1 k_2 [14]}{k_{-1} + k_2[14]} \tag{13}$$

10. (cont.) and since $-d[13]/[13] = -d\ln[13]$, equation (13) can now be written as

$$\frac{-d\ln[13]}{dt} = \frac{k_1 k_2 [14]}{k_{-1} + k_2 [14]} \tag{14}$$

The physical significance of this equation is that $-d\ln[13]/dt$ is simply k_{obs} and is equal to the slope of the line arising from a plot of $\ln[13]$ vs time. A plot of $1/k_{obs}$ vs $1/[14]$ should be linear with a slope of $k_{-1}/k_1 k_2$, as given by equation (15).

$$\frac{1}{k_{obs}} = \frac{k_{-1}}{k_1 k_2 [14]} + \frac{1}{k_1} \tag{15}$$

The fact that completely different products are observed when benzene was used as a solvent is a result of the fact that CO loss followed by alkane elimination (Mechanism 3) is faster than alkyl migration in non-stabilizing solvents.

Reaction 2

The possible reaction pathways that yield alkanes as products are 1), 3), and 4).

1)

$$M\Big\langle{}^{CO}_{R} \longrightarrow M-\overset{\overset{\displaystyle O}{\|}}{C}-R \overset{M'-H}{\longrightarrow} M-CO \;+\; R-H \;+\; M'$$

3)

$$M\Big\langle{}^{CO}_{R} \overset{-CO}{\longrightarrow} M-R \overset{M'-H}{\longrightarrow} R-H \;+\; M' \;+\; M$$

4)

$$M-R \longrightarrow M\cdot \;+\; R\cdot \overset{M'-H}{\longrightarrow} R-H \;+\; M'$$

The fact that crossover products are observed for the reaction of **17** with $Mn_2(CO)_8(PMePh_2)_2$ implies the presence of a radical species. This is also supported by observation 2), which suggests that $\cdot Mn(CO)_4 P$ is present and reacts rapidly with **14** to give $H-Mn(CO)_4 P$. The postulated mechanism is shown on the following page:

10. (cont.)

$$\underline{17} \xrightarrow[k_{-3}]{k_3} \quad p-MeOC_6H_4CH_2\cdot \quad + \quad Mn(CO)_4P\cdot$$

$$p-MeOC_6H_4CH_2\cdot \; + \; \underline{14} \xrightarrow{\text{fast}} p-MeOC_6H_4CH_3 \; + \; \cdot Mn(CO)_5$$
$$\underline{21}$$

$$Mn(CO)_4P\cdot \; + \; 14 \longrightarrow H-Mn(CO)_4P \; + \; \cdot Mn(CO)_5$$
$$\underline{19}$$

$$2 \; \cdot Mn(CO)_5 \longrightarrow Mn_2(CO)_{10}$$
$$\underline{18}$$

$$\cdot Mn(CO)_5 \; + \; \cdot Mn(CO)_4P \longrightarrow Mn_2(CO)_9P$$
$$\underline{20}$$

The fact that **17** does not lose CO at the reaction temperature suggests that mechanism 3) is not important for this system. This observation is consistent with a stronger M–CO bond expected for a complex with better backbonding capabilities such as **17**.

Nappa, M.J.; Santi, R.; Halpern, J. Organomet. **1985**, _4_, 34.

11. The allyl moieties in the following products can be introduced into the molecule either from the allyl chloride or from the allyl tin reagent; however, since the allyl tin reagents are made via a Grignard reaction, it is typically easier to have more of the functionality on the allyl chloride.

a) This compound can be made from either route.

11.   (cont.)

b)

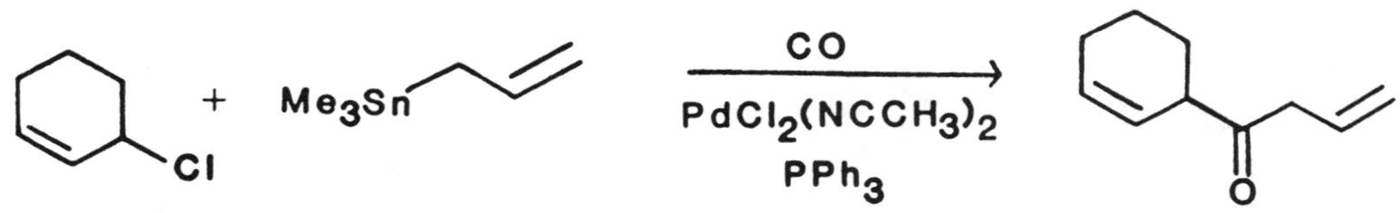

c)

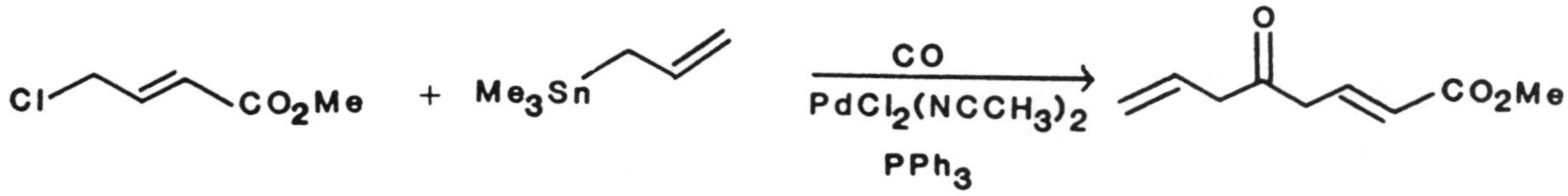

Merrifield, J.H.; Godschalx, J.P.; Stille, J.K. _Organomet._   1984, _3_, 1108.

Merrifield, J.H., private communication.

11

Catalytic Reactions

QUESTIONS

1. The following catalytic reaction is observed (L = P(O-o-tolyl)$_3$):

$$L_2Ni(C_2H_4) \quad + \quad HCN \quad + \quad C_2H_4 \quad \longrightarrow \quad L_2Ni(C_2H_4) \quad + \quad EtCN$$

$$\underline{\textbf{1}} \qquad\qquad\qquad\qquad\qquad\qquad\qquad\qquad \underline{\textbf{1}}$$

OBSERVATIONS

1) $\Delta H^{\ddagger} = 8.9$ kcal/mol; $\Delta S^{\ddagger} = -34$ eu

2) During the reaction, intermediate isomeric species **2a-c** can be detected by phosphorous, proton, and carbon NMR spectroscopies. The observed data for the reaction mixture are given below.

^{31}P NMR (δ, ppm):

116.9
117.7
118.1
129.8   (free P(O-o-tolyl)$_3$

^{13}C NMR using ^{13}C-labeled C$_2$H$_4$ (δ, ppm):

11.7
14.1
58.9

^{1}H NMR (δ, ppm):

0.61 (m)
2.03 (s)
2.09 (br, s)

3) The rate law is of the form d[EtCN]/dt = k[**2**][L].

a) Identify **2a-c**.

b) Propose a stepwise mechanism for the reaction and verify that your mechanism agrees with the observed rate law.

2. Propose a mechanism(s) for the following reactions.

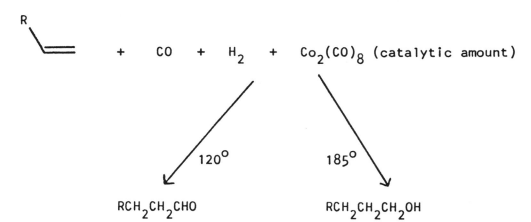

RCH$_2$CH$_2$CHO

RCH$_2$CH$_2$CH$_2$OH

3. The complex $(\eta^5-C_5H_5)Ni(\eta^3-C_5H_7)$, **4**, catalyzes the conversion of ethylene to 1-butene. Propose a catalytic cycle for the conversion of ethylene to 1-butene which is consistent with the following observations:

OBSERVATIONS
1) Introduction of a solution of **4** into a gas chromatography column at 150°C indicates the presence of both cyclopentadiene and cyclopentadiene dimer.

2) If the reaction is carried out in the presence of one equivalent of triphenylphosphine, 1-butene still accounts for most of the product; however, the percentage conversion of ethylene to 1-butene is greatly decreased, even if the reaction is allowed to proceed for longer times. Complex **5**, CpNi(PPh$_3$)(n-C$_4$H$_9$), is isolated from this reaction mixture.

HINT
Consider a Ni-H species as an intermediate.

4. Propose a mechanism for the following conversion. Be sure your mechanism accounts for an induction period which is observed for this reaction.

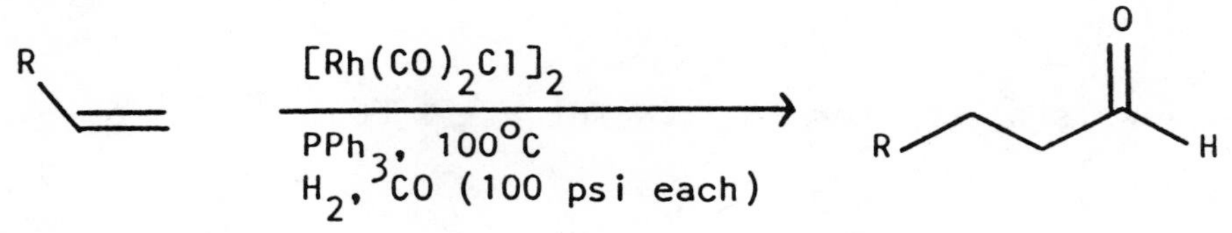

5. Propose a mechanism for the following metathesis reaction.

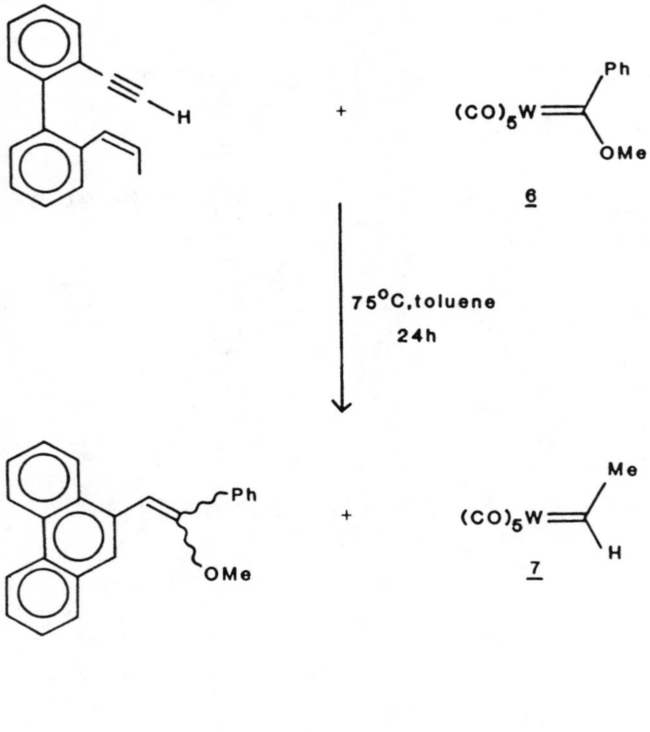

6. Benzyl bromide can be converted to methylphenylacetate in one pot, as shown below.

$$PhCH_2Br \quad + \quad CO \quad + \quad KOMe \quad \xrightarrow{\quad Fe(CO)_5 \quad} \quad PhCH_2COOMe$$

Propose a mechanism for this reaction which is catalytic in $Fe(CO)_5$ and stoichiometric in the other reagents.

7. Propose structures for compounds **8** and **10**.

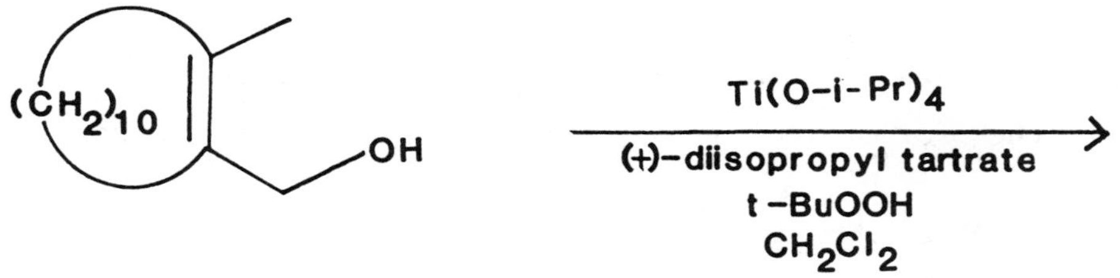

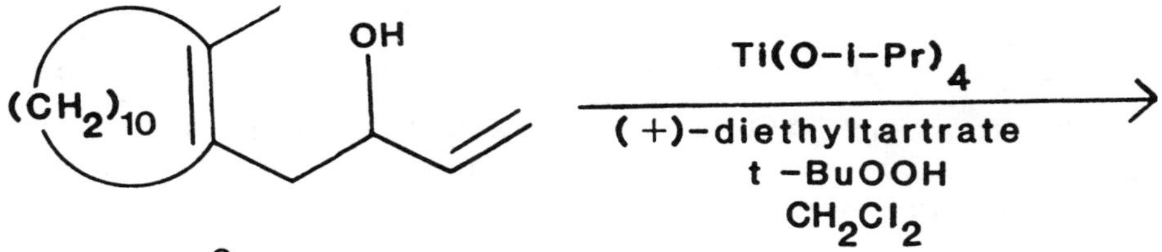

ANSWERS

1. a) Intermediates **2a–c** are shown below.

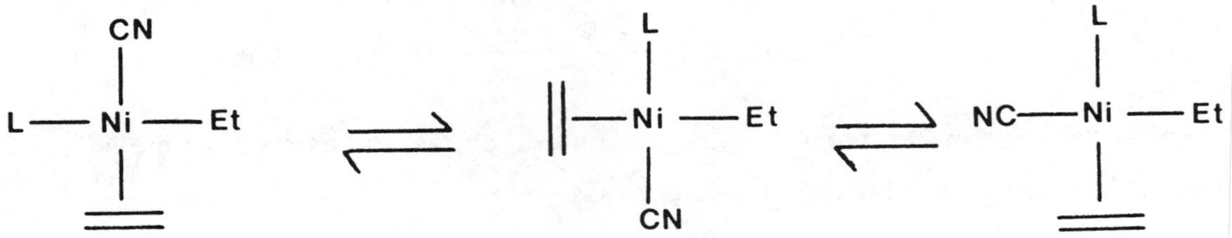

$$
\underset{\textbf{2a}}{L-Ni-Et} \;\; \rightleftharpoons \;\; \underset{\textbf{2b}}{Ni-Et} \;\; \rightleftharpoons \;\; \underset{\textbf{2c}}{NC-Ni-Et}
$$

Complex **2a** is the predominant isomer, but not necessarily the most reactive.

b) In the scheme below, L = P(O–o–tolyl)$_3$. Three pieces of information suggest that the formation of **3** from **2** is the slow step in the reaction:

1) The observation of the stable intermediates **2a–c**,
2) the dependence of the rate on both [**2**] and L, and
3) the large negative entropy of activation, which supports an associative transition state.

The stability of the nickel complex, L$_2$Ni(C$_2$H$_4$), **1**, produced upon reductive–elimination of Et–CN is certainly a driving force for reductive–elimination to occur from the 5-coordinate nickel species **3**, as opposed to reductive–elimination occurring directly from the 4-coordinate species **2**.

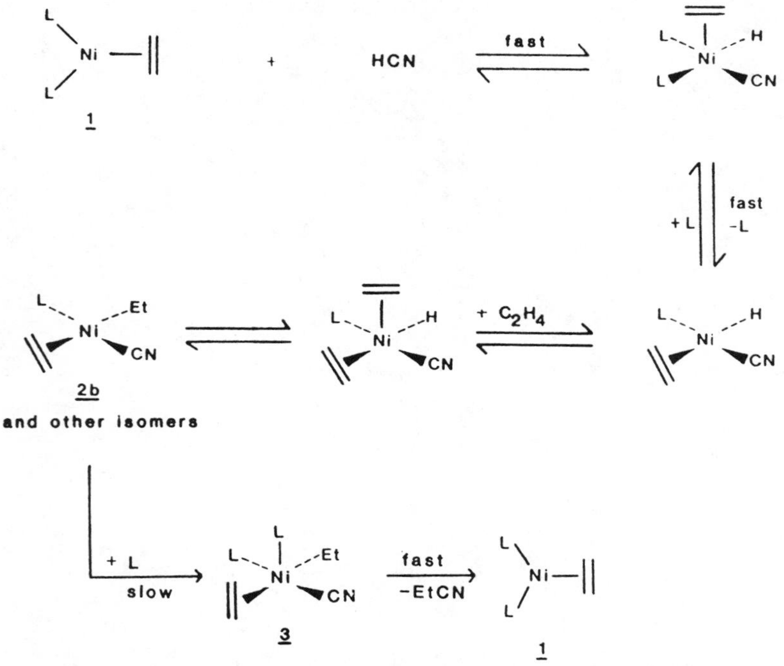

McKinney, R.J.; Roe, D.C. *J. Am. Chem. Soc.* **1985**, *107*, 261.

2. At 120°C, the reaction proceeds as shown below:

$$Co_2(CO)_8 \ + \ H_2 \ \rightleftharpoons \ 2 \ HCo(CO)_4$$

$$HCo(CO)_4 \ \rightleftharpoons \ HCo(CO)_3 \ + \ CO$$

$$HCo(CO)_3 \ + \ CH_2{=}CHR \ \rightleftharpoons \ HCo(CO)_3(CH_2{=}CHR)$$

$$HCo(CO)_3(CH_2{=}CHR) \ \rightleftharpoons \ RCH_2CH_2Co(CO)_3$$

$$RCH_2CH_2Co(CO)_3 \ \xrightarrow{CO} \ RCH_2CH_2\overset{\overset{\displaystyle O}{\|}}{C}{-}Co(CO)_3$$

$$RCH_2CH_2\overset{\overset{\displaystyle O}{\|}}{C}{-}Co(CO)_3 \ + \ H_2 \ \rightleftharpoons \ RCH_2CH_2\overset{\overset{\displaystyle O}{\|}}{\underset{\underset{\displaystyle H}{|}}{C}}{-}\overset{\overset{\displaystyle H}{|}}{Co}(CO)_3$$

$$RCH_2CH_2\overset{\overset{\displaystyle O}{\|}}{\underset{\underset{\displaystyle H}{|}}{C}}{-}Co(CO)_3 \ \longrightarrow \ RCH_2CH_2\overset{\overset{\displaystyle O}{\|}}{C}{-}H \ + \ HCo(CO)_3$$

Note that the last step produces $HCo(CO)_3$ which can proceed to react with additional olefin and thus repeat the cycle.

At higher temperatures, reduction of the aldehyde with $HCo(CO)_4$ is competitive with the process shown above, and the alcohol is observed as the major product of the reaction.

$$RCH_2CH_2\overset{\overset{\displaystyle O}{\|}}{C}{-}H \ + \ HCo(CO)_3 \ \rightleftharpoons \ RCH_2CH_2CH_2{-}O{-}Co(CO)_3$$

$$RCH_2CH_2CH_2{-}O{-}Co(CO)_3 \ + \ H_2 \ \rightleftharpoons \ RCH_2CH_2CH_2{-}O{-}\overset{\overset{\displaystyle H}{|}}{\underset{\underset{\displaystyle H}{|}}{Co}}(CO)_3$$

$$RCH_2CH_2CH_2{-}O{-}\overset{\overset{\displaystyle H}{|}}{Co}(CO)_3 \ \longrightarrow \ RCH_2CH_2CH_2{-}OH \ + \ HCo(CO)_3$$

Orchin, M. Accts. Chem. Res. 1981, 14, 259.

3. The proposed mechanism is shown below:

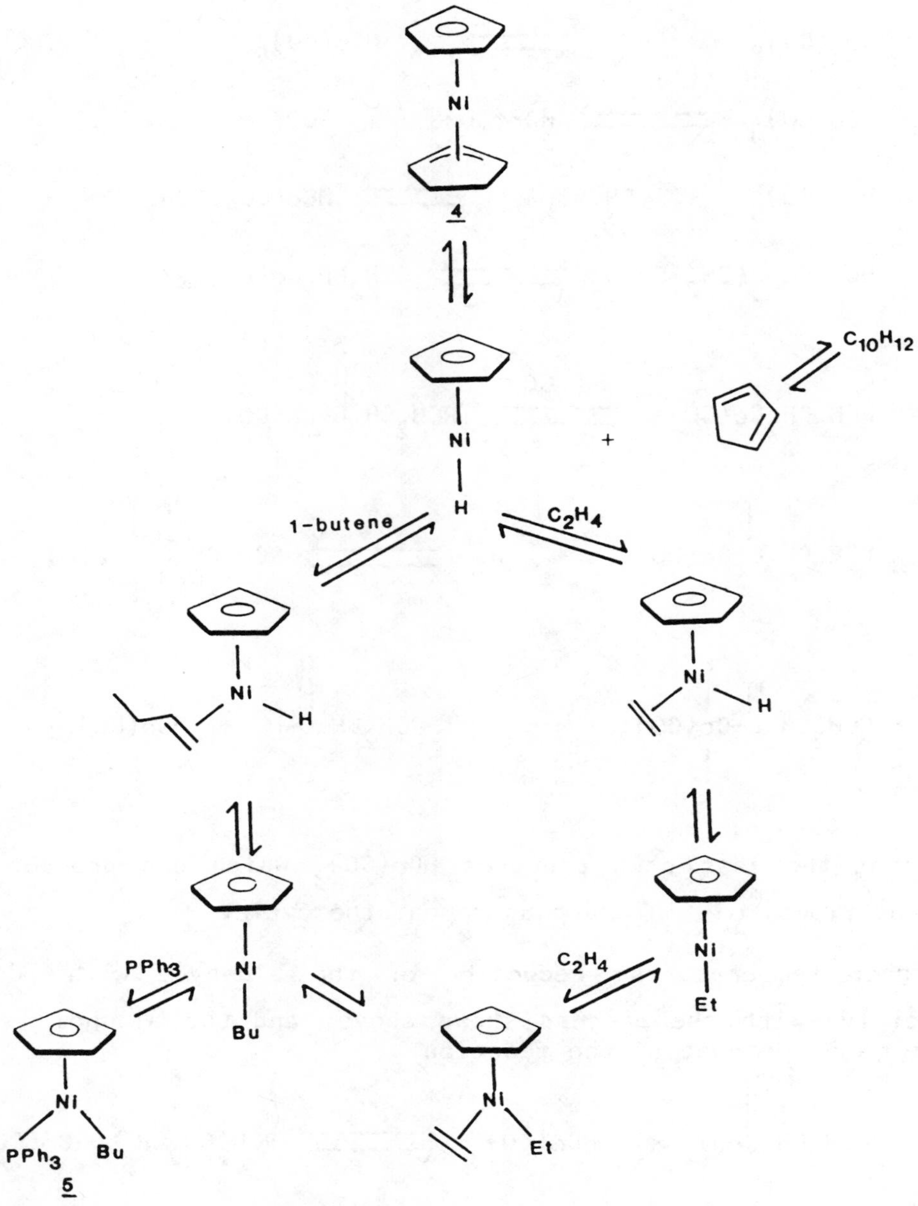

An initial beta-hydrogen abstraction from the η^3-C_5H_7 ring gives free cyclopentadiene (which rapidly dimerizes when heated) and a CpNi-H species that adds a molecule of ethylene. Olefin insertion to give CpNi-Et followed by addition of another molecule of ethylene and another insertion reaction gives the butyl complex, CpNi-Bu. Beta-hydrogen abstraction, followed by ligand dissociation gives the observed 1-butene and regenerates the catalyst CpNi-H. Addition of PPh$_3$ to the reaction decreases the amount of catalyst in solution by forming complex 5.

McClure, J.D.; Barnett, K.W. J. Organomet. Chem. 1974, 80, 385.

4. In the scheme below, L = PPh$_3$.

The active catalyst is complex **11**. The induction period is the time necessary to form this complex.

Pruett, R.L. Adv. Organomet. Chem. **1979, 17**, 1.

Evans, D.; Osborn, J.A.; Wilkinson, G. J. Chem. Soc. A **1968**, 3133.

5. This is not a simple olefin metathesis reaction, since a third ring has been formed. Schematically, the proposed mechanism is:

5. (cont.) Since a carbene is one of the products, this type of reaction can be performed catalytically using the heteroatomcarbene, **6**, as an initiator. The propagation step for the catalytic reaction is:

Katz, T.J.; Sivavec, T.M. J. Am. Chem. Soc. 1985, 107, 737.

6. The proposed mechanism is:

$$KOMe \;+\; Fe(CO)_5 \longrightarrow [(CO)_4Fe-COOMe]^- K^+$$

$$[(CO)_4Fe-COOMe]^- K^+ \;+\; PhCH_2Br \longrightarrow (CO)_4Fe-COOMe$$
$$\qquad\qquad\qquad\qquad\qquad\qquad\qquad\qquad\qquad\qquad\qquad |$$
$$\qquad\qquad\qquad\qquad\qquad\qquad\qquad\qquad\qquad\qquad CH_2Ph$$

$$(CO)_4Fe-COOMe \longrightarrow PhCH_2COOMe \;+\; Fe(CO)_4$$
$$\;\;\; |$$
$$\;\;\; CH_2Ph$$

$$Fe(CO)_4 \;+\; CO \longrightarrow Fe(CO)_5$$

Tustin, G.C.; Hembre, R.T. J. Org. Chem. 1984, 49, 1761.

7. The first reaction is a variation on a scheme developed by K.B. Sharpless for the conversion of allyl alcohols to epoxy-alcohols. In this particular example, the (−)-epoxy-alcohol (**8**) is obtained in 94% enantiomeric excess. The catalyst contains the optically-active tartrate ligand and is the source of the stereospecificity of the reaction.

7. (cont.) For the second reaction, the epoxide formed is the homoallylic epoxide, **10**, rather than the allyl epoxide formed in the previous example. This is probably due to the greater nucleophilicity (greater substitution) of the homoallyllic double bond. In general, homoallylic epoxidations proceed in the opposite absolute sense, with regard to stereospecificity, compared to the allylic epoxidations.

(R,S)- **9**   (S)- **10**

Marshall, J.A.; Jenson, T.M. J. Org. Chem. **1984**, 49, 1707.

Katsuki, T.; Sharpless, K.B. J. Am. Chem. Soc. **1980**, 102, 5974.

Martin, V.S.; Woodard, S.S.; Katsuki, T.; Yamada, Y.; Ikeda, M.; Sharpless, K.B. J. Am. Chem. Soc. **1981**, 103, 6237.

12

Organometallic Complexes in Organic Synthesis

QUESTIONS

Issue number 24 of <u>Tetrahedron</u> for **1985** contains a large number of papers dealing with applications of transition metal organometallic chemistry to organic synthesis.

1. Provide a structure for compounds **1**, **2**, **3**, and **4**.

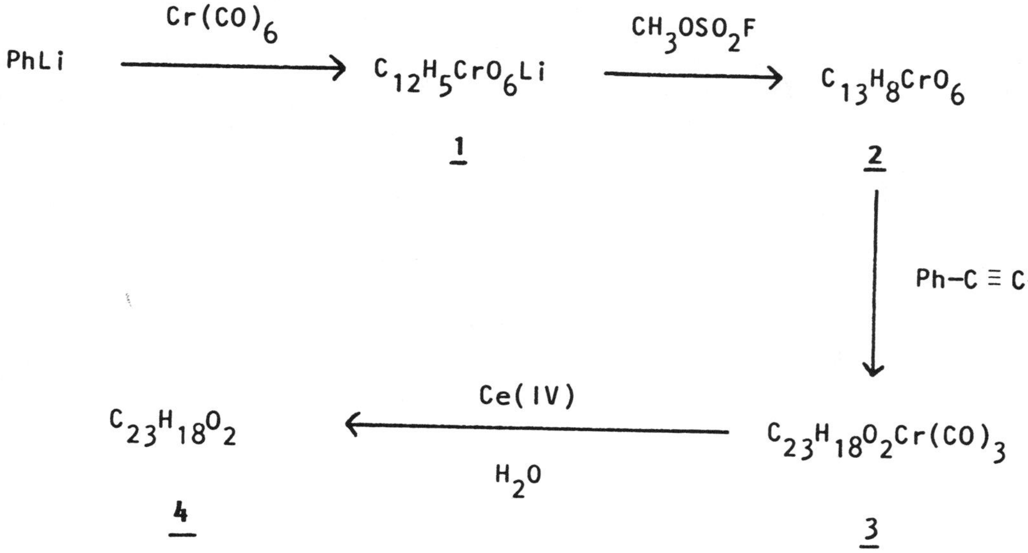

PhLi
Cr(CO)6
C12H5CrO6Li
1
CH3OSO2F
C13H8CrO6
2
Ph–C ≡ C–Ph
C23H18O2Cr(CO)3
3
Ce(IV)
H2O
C23H18O2
4

Compound **4** is soluble in NaOH. Its NMR spectrum consists of a singlet at 3.2 ppm and aromatic resonances, and its IR spectrum has a peak at 3550 cm^{-1}.

2. Propose a mechanism for the following transformation:

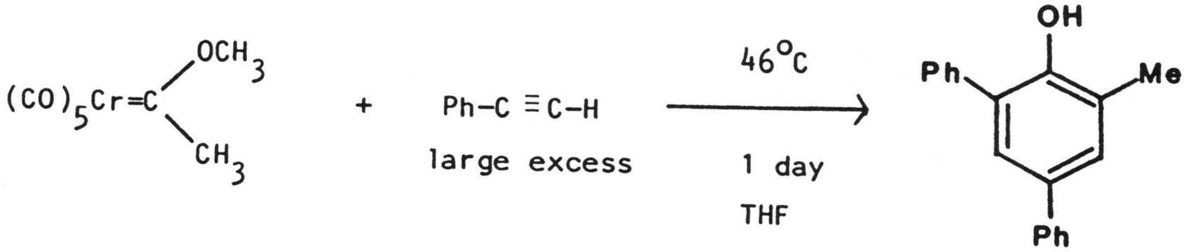

(CO)5Cr=C
OCH3
CH3
+
Ph–C ≡ C–H
large excess
46°C
1 day
THF
OH
Ph
Me
Ph

3. Predict the products and propose a mechanism for the following reaction.

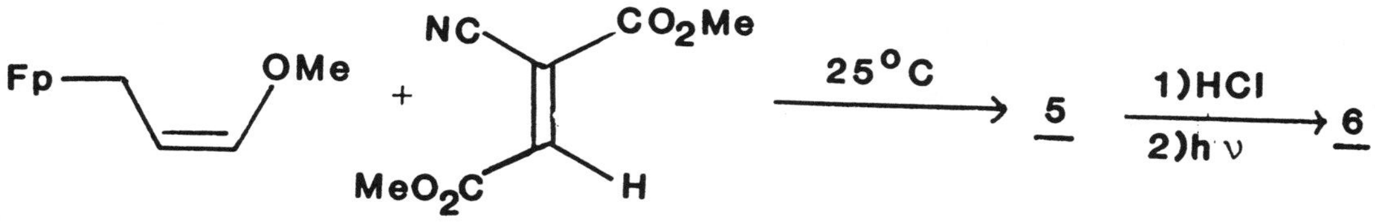

Fp
OMe
+
NC
CO2Me
MeO2C
H
25°C
5
1)HCl
2)hν
6

4. Propose structures for compounds **7** and **8** and assign all spectral data.

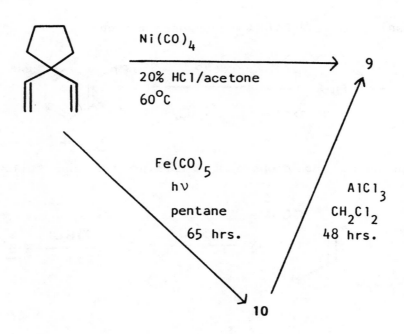

Spectral data for **7**

^{1}H NMR (δ, CDCl$_3$)

0.17 (s, 9H)
0.71 (t, J=7 Hz, 3 H)
0.92–1.44 (m, 2 H)
1.64–1.98 (m, 2 H)
2.99 (br d, J=4.7 Hz, 2 H)
3.53 (s, 3 H)
4.65 (dd, J=6.2, 6.2 Hz, 1 H)
5.08–5.26 (m, 4 H)

IR (neat, cm^{-1})

1960 (s), 1886 (s)

Spectral data for **8**

^{1}H NMR (δ, CDCl$_3$)

0.85 (t, 3 H)
1.12–1.56 (m, 2 H)
1.71–2.24 (m, 6 H)
3.20 (dd, J=14.4, 3.8 Hz, 1 H)
3.48 (dd, J=14.4, 4.3 Hz, 1 H)
3.80 (s, 3 H)
4.03 (dd, J=8.6, 5.6 Hz, 1 H)
4.96–6.02 (m, 5 H)
6.74–7.36 (m, 3 H)

5. Give structures for compounds **9** and **10** and assign all spectral data.

5. (cont.)

Spectral data for **9**

IR (cm^{-1})
_____

1740

^{1}H NMR (δ, CCl$_4$)
_____
0.93 (d, J=7 Hz, 3 H)
1.23–2.30 (m, 13 H)

Mass Spectrum (m/e)
_____

152 (38%)

Spectral data for **10**

IR (cm^{-1})
_____

2035, 1970–1950

^{1}H NMR (δ, C$_6$D$_6$)
_____
0.33(dd, J=12, 2 Hz, 2 H)
0.50–0.70 (m, 8 H)
1.07 (dd, J=7, 2 Hz, 2 H)
2.68 (dd, J=12, 7 Hz, 2H)

Mass Spectrum (m/e)
_____

262 (7%), 234 (17%),
206 (23%), 178 (100%)

6. As shown below, the reaction of the cationic carbene complex Fp=C(H)CH$_3^+$ with <u>para</u>-substituted styrenes generates aryl methyl cyclopropanes. The following relative rates were obtained:

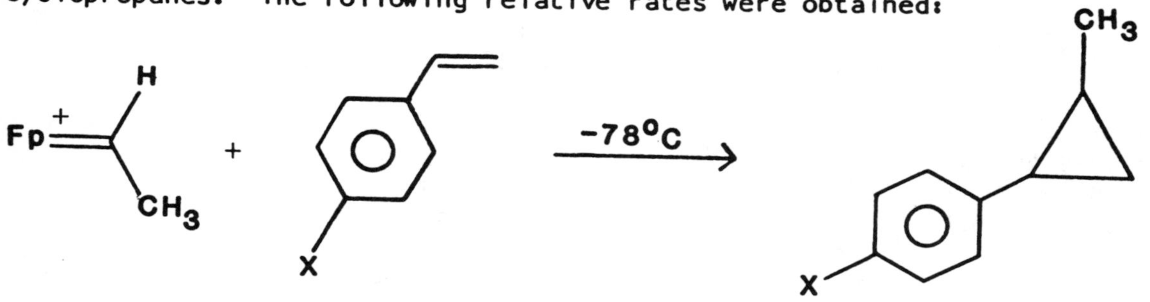

cis and trans

X	k$_{rel}$	σ_p^+
OMe	74	−0.78
Me	5.9	−0.31
H	1.0	0.00
F	0.9	−0.07
Cl	0.3	0.11

a) Using a Hammett plot of the equation, $\log k_{rel} = \sigma_p^+ \rho$, find the rho ($\rho$) value for this reaction.

6.   b) What does this value of rho indicate about the nature of the transition state?

c) Propose a mechanism for this reaction.

7.   Propose structures for compounds **11**, **12**, and **13**.

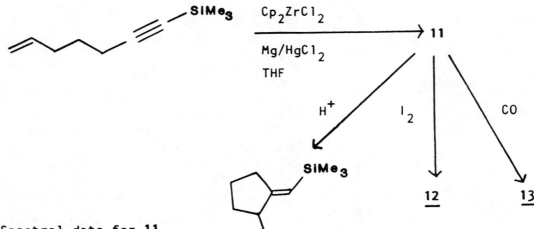

Spectral data for **11**

^{1}H NMR (δ, toluene-d_8)

0.11 (s, 9 H)
0.90–2.40 (m, 9 H)
5.78 (s, 5 H)
5.82 (s, 5 H)

^{13}C{^{1}H} (δ, toluene-d_8)

1.75
20.89
33.63
38.50
41.51
43.09
109.31
110.45
152.10
156.32

8.   Consider the following reaction, first reported by Pauson and co-workers (J. Chem. Soc., Perkin I 1973, 977; 1976, 30), in which "L" represents the larger of the two groups of the coordinated acetylene and "S" represents the smaller group. Complexes like **14** can be made by reaction of an acetylene with $Co_2(CO)_8$.

8. (cont.) Based on this reactivity, propose a structure for the product of each of the following reactions:

a)

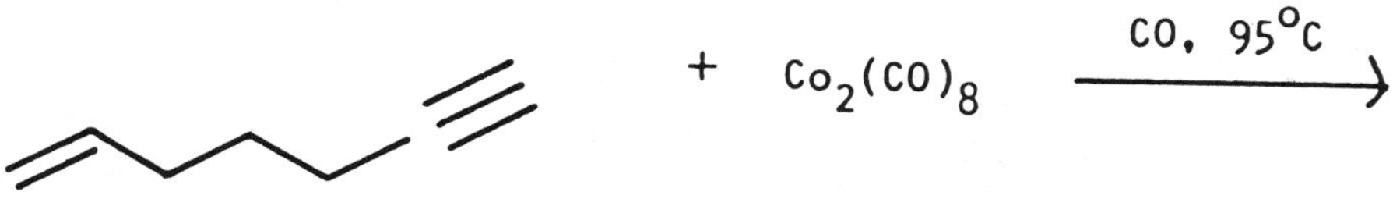

$+$ $Co_2(CO)_8$ $\xrightarrow{\text{CO, 95°C}}$

b)

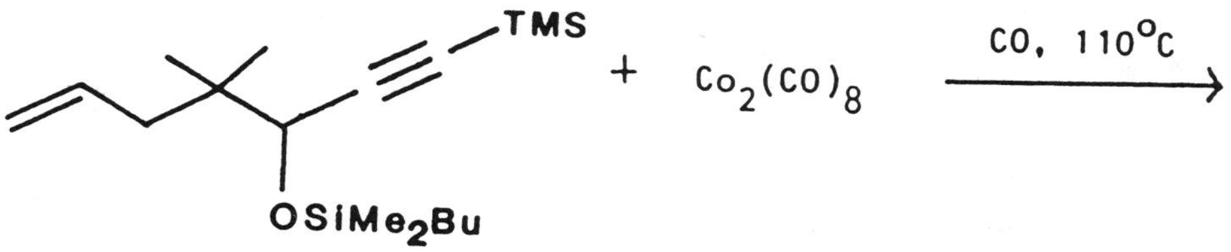

$+$ $Co_2(CO)_8$ $\xrightarrow{\text{CO, 110°C}}$

c)

$+$ $Co_2(CO)_8$ $\xrightarrow{\text{CO, 60°C}}$

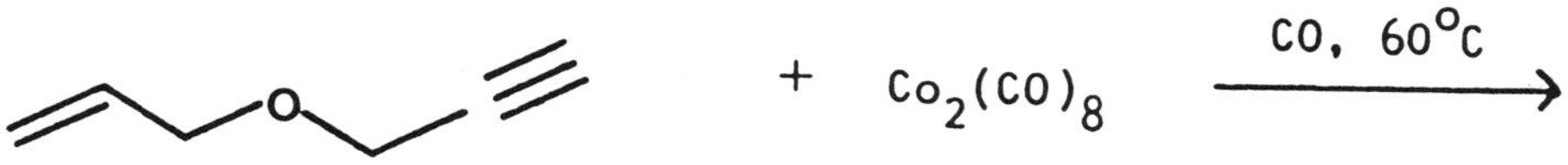

9. Starting with $Cr(CO)_6$, design a synthesis of compound **18**. In addition, propose a structure for **19** and give its mechanism of formation from **18**.

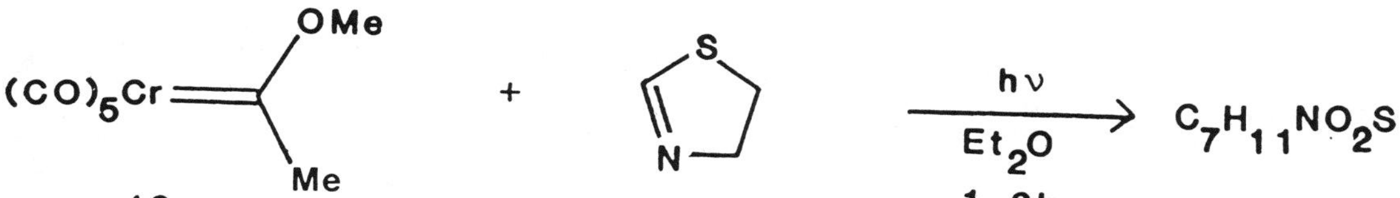

$(CO)_5Cr=$ OMe / Me

18

$+$

$\xrightarrow[\text{Et}_2\text{O}]{h\nu}$ $\xrightarrow{1-3h}$ $C_7H_{11}NO_2S$

19

Spectral data for **19**

IR (cm^{-1})	^{1}H NMR (δ, $CDCl_3$)
1770	1.38 (s, 3 H)
	2.76 (m, 1 H)
	2.85 (m, 1 H)
	3.06 (m, 1 H)
	4.18 (m, 1 H)
	3.48 (s, 3 H)
	5.02 (s, 1 H)

10. Propose structures for compounds **20** and **21**.

HINT
Compound **21** is related to compound **19** of the previous problem.

Spectral data for **20**

IR (cm^{-1})	^{1}H NMR (δ, CDCl$_3$)
1918	0.79 (t, J=7 Hz, 3 H)
1591	1.36 (m, 2 H)
	2.16 (m, 2 H)
	2.84 (dd, J=18, 2 Hz, 1 H)
	3.26 (dd, J=18, 10 Hz, 1 H)
	3.47 (dd, J=10, 2 Hz, 1 H)
	4.41 (d, J=1 Hz, 5 H)
	7.07-7.73 (m, 15 H)

11. Propose a structure for compound **22** and a mechanism for its formation.

$$Ni(CO)_4 \xrightarrow[\text{2) 90 min.,}]{\text{1) n-PrLi, -50}^\circ\text{C, 30 min.}} \underline{22}$$

3) allyl iodide/overnight,
room temperature

11. (cont.)
Spectral data for **22**

IR (CHCl$_3$, cm^{-1})	^{1}H NMR (δ, CDCl$_3$)
3080 (w)	0.90 (t, J=7.0 Hz, 3 H)
3000 (m)	1.4–1.7 (m, 2 H)
2970 (s)	2.2–3.2 (m, 5 H)
2840 (w)	3.65 (d, J=6.5 Hz, 1 H)
1710 (s)	3.88 (s, 3 H)
1690 (s)	4.0–4.5 (m, 4 H)
1640 (w)	4.9–5.2 (m, 2 H)
1590 (s)	5.6–6.0 (m, 1 H)
1480 (s)	7.13 (dd, J=8.0, 1.5 Hz, 1 H)
1270 (s)	7.40 (t, J=7.0 Hz, 1 H)
	7.70 (dd, J=7.0, 2.0 Hz, 1 H)

ANSWERS

1. Compounds **1–4** are shown below:

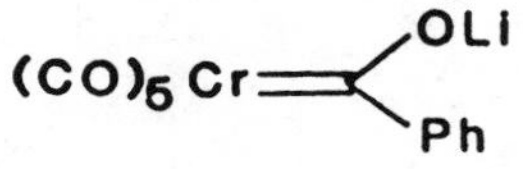

$$\underline{1} \qquad \underline{2}$$

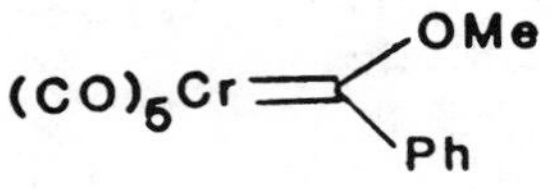

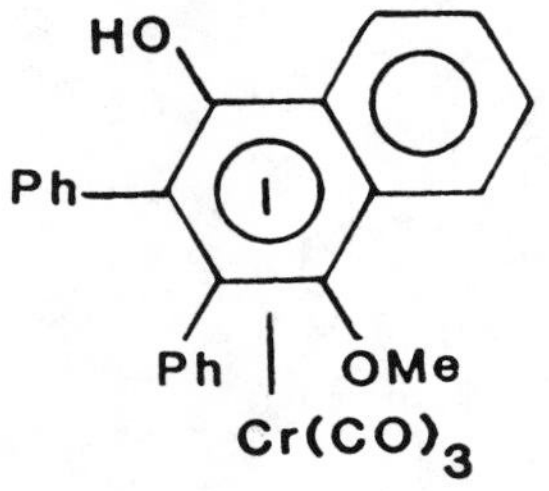

$$\underline{3} \qquad \underline{4}$$

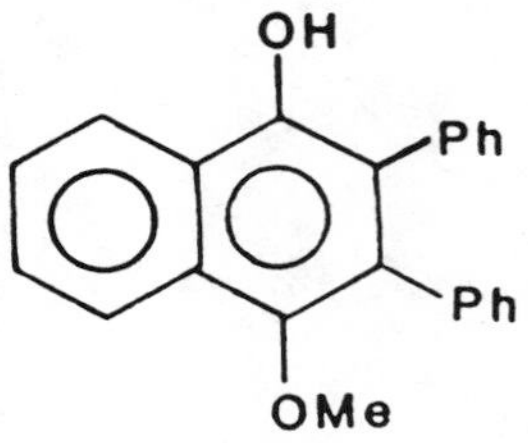

Dotz, K.H. Angew. Chem. Int. Ed. Engl. **1975**, <u>14</u>, 644.

For a more recent application of this reaction, see: Wulff, W.D.; Tang, P.-C. J. Am. Chem. Soc. **1984**, <u>106</u>, 434.

2. The proposed mechanism is shown below:

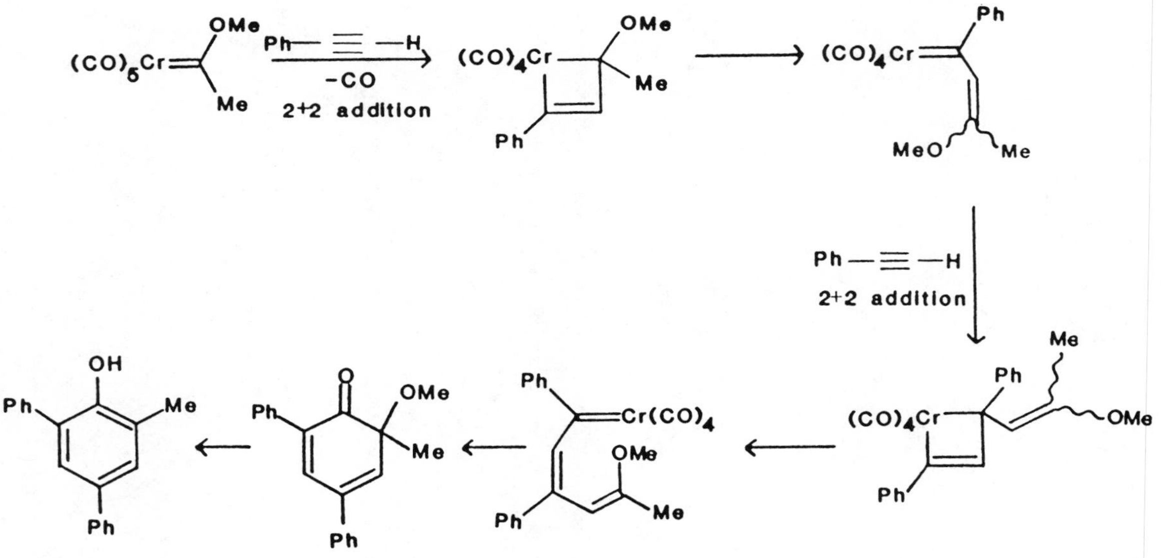

Under the reaction conditions, the ketone is reduced to the phenol.

Wulff, W.D.; Kaesler, R.W.; Peterson, G.A.; Tang, P.-C. J. Am. Chem. Soc. **1985**, <u>107</u>, 1060.

3. The proposed mechanism is shown below:

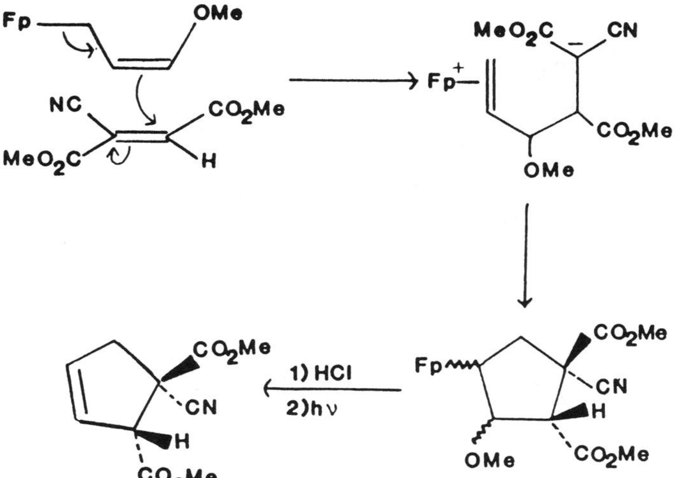

Baker, R.; Keen, R.B.; Morris, M.D.; Turner, R.W. J. Chem. Soc., Chem. Comm. **1984**, 987.

4. Both steps involve a nucleophilic attack on the coordinated benzene ring. I_2 serves to oxidatively cleave the product from the metal. Complexes **7** and **8** are:

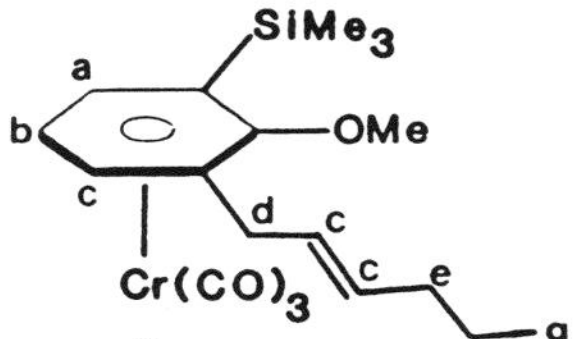

The NMR data are assigned as follows:

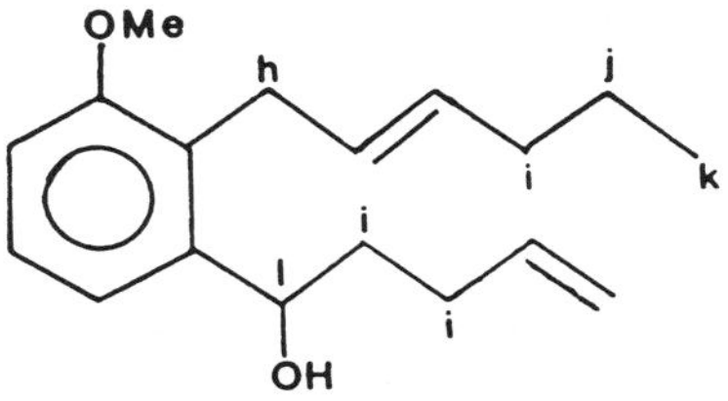

Spectral data for **7**

^{1}H NMR (δ, CDCl$_3$)

0.17 (s, 9H, SiMe$_3$)

0.71 (t, J=7 Hz, 3 H, H$_g$)

0.92–1.44 (m, 2 H, H$_f$)

1.64–1.98 (m, 2 H, H$_e$)

2.99 (br d, J=4.7 Hz, 2 H, H$_d$)

3.53 (s, 3 H, OMe)

4.65 (dd, J=6.2, 6.2 Hz, 1 H, H$_b$)

5.08–5.26 (m, 4 H$_a$, H$_c$)

Spectral data for **8**

^{1}H NMR (δ, CDCl$_3$)

0.85 (t, 3 H, H$_k$)

1.12–1.56 (m, 2 H, H$_j$)

1.71–2.24 (m, 6 H, H$_i$)

3.20 (dd, J=14.4, 3.8 Hz, 1 H, H$_h$)

3.48 (dd, J=14.4, 4.3 Hz, 1 H, H$_{h'}$)

3.80 (s, 3 H, OMe)

4.03 (dd, J=8.6, 5.6 Hz, 1 H, H$_l$)

4.96–6.02 (m, 5 H, vinyls)

6.74–7.36 (m, 3 H, aromatics)

4. (cont.)

IR for 7 (neat, cm^{-1})

1960 (s), 1886 (s): Terminal metal carbonyl stretches.

Semmelhack, M.F.; Zask, A. J. Am. Chem. Soc. 1983, 105, 2034.

5. Complexes 9 and 10 are shown below:

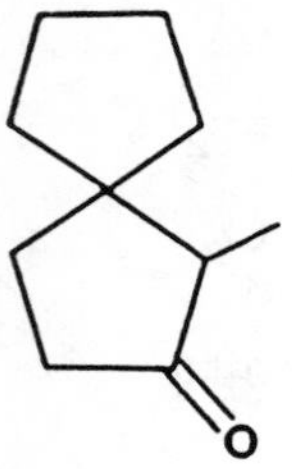

The spectral data are assigned as follows:

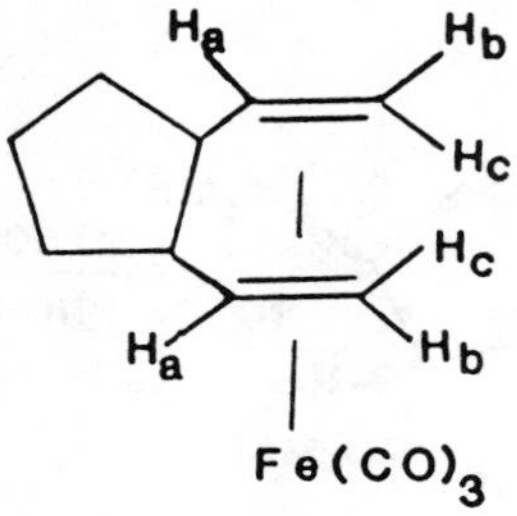

Spectral data for 9	Spectral data for 10
IR (cm^{-1})	IR (cm^{-1})
1740 (C=O of 5-membered ring)	2035, 1970–1950 (terminal M–C≡O's)
^{1}H NMR (δ, CCl$_4$)	^{1}H NMR (δ, C$_6$D$_6$)
0.93 (d, J=7 Hz, 3 H, Me)	0.33(dd, J=12, 2 Hz, 2 H, H$_c$)
1.23–2.30 (m, 13 H, methylenes)	0.50–0.70 (m, 8 H, methylenes)
	1.07 (dd, J=7, 2 Hz, 2 H, H$_b$)
	2.68 (dd, J=12, 7 Hz, 2H, H$_a$)
Mass Spectrum (m/e)	Mass Spectrum (m/e)
152 (38%), M$^+$	262 (7%), M$^+$
	234 (17%), M$^+$–CO
	206 (23%), M$^+$–2 CO
	178 (100%), M$^+$–3 CO

Eilbracht, P.; Acker, M.; Totzauer, W. Chem. Ber. 1983, 116, 238.

6. a) The Hammett plot is shown below:

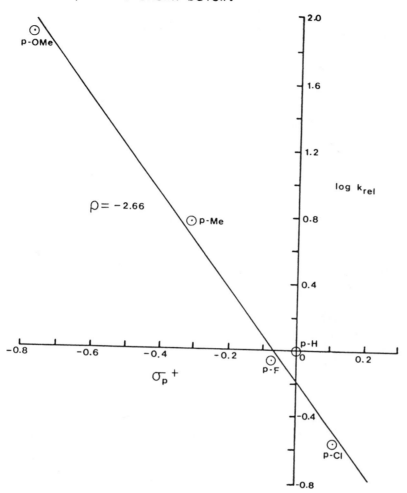

Sigma <u>plus</u> substituent constants generally give a better correlation than sigma constants for reactions in which the substituent can enter into direct resonance interaction with an electron-deficient reaction site in the transition state. The reaction constant rho is the slope of the line and is equal to -2.66 for this reaction.

b) The rho value for a reaction gives a measure of the amount of charge development at the carbon alpha to the aromatic ring in the transition state. A negative rho value indicates that the reaction is accelerated by electron-donating substituents and is supportive of a highly polarized cationic transition state, but not one with sufficient charge development at the alpha carbon such that it could be considered a free carbonium ion. (For comparison, consider the rho value of -4.48 obtained for the solvolysis of cumyl chloride $(C_6H_5CMe_2Cl)$, in which the transition state closely resembles a free carbonium ion.) The linearity of these points indicates that no change in mechanism is occurring with a change in the electron-donating ability of the substituent.

c) The proposed mechanism is shown on the following page:

6. (cont.)

cis cyclopropanes   trans cyclopropanes

Brookhart, M.; Kegley, S.E.; Husk, G.R. Organomet. **1984**, <u>3</u>, 650.

Kegley, S., Ph.D. Thesis, University of North Carolina, 1982.

For more information on Hammett relationships, see March, J. "Advanced Organic Chemistry, Reactions, Mechanisms, and Structure", 3rd ed., John Wiley and Sons (New York, 1985), p. 242.

7. The structures of **11**, **12**, and **13** are shown below:

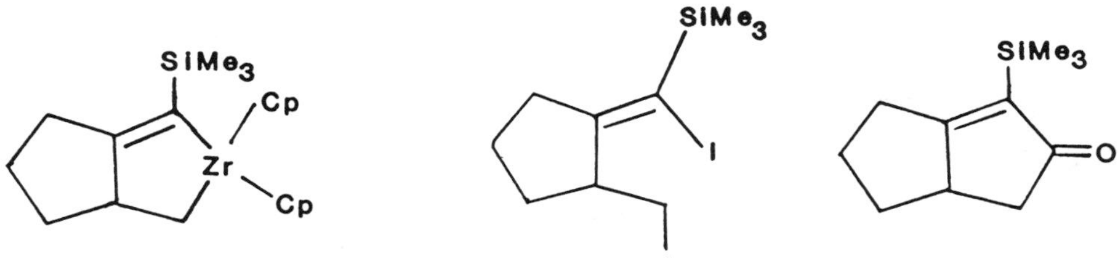

11   **12**   **13**

Negishi, E.; Holmes, S.J.; Tour, J.M.; Miller, J.A. **J.** **Am.** **Chem.** **Soc.** **1985**, _107_, 2568.

8. At room temperature, the cobalt-acetylene complex is formed. Upon heating, this complex reacts with the olefin and CO to give the cyclopentenone derivative.

a)

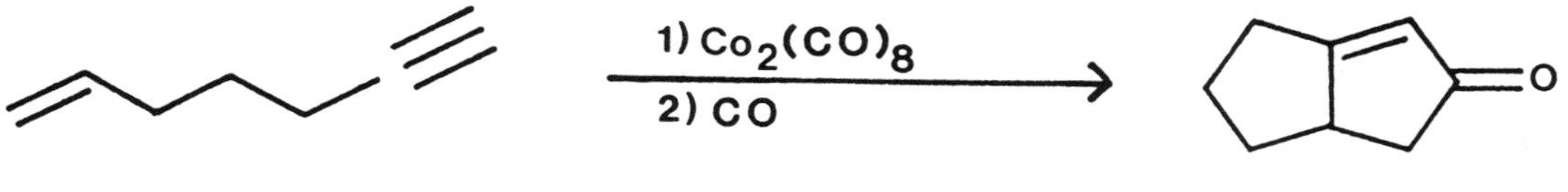

1) $Co_2(CO)_8$
2) CO

30% yield

Schore, N.E.; Croudace, M.C. **J.** **Org.** **Chem.** **1981**, _46_, 5436.

b)

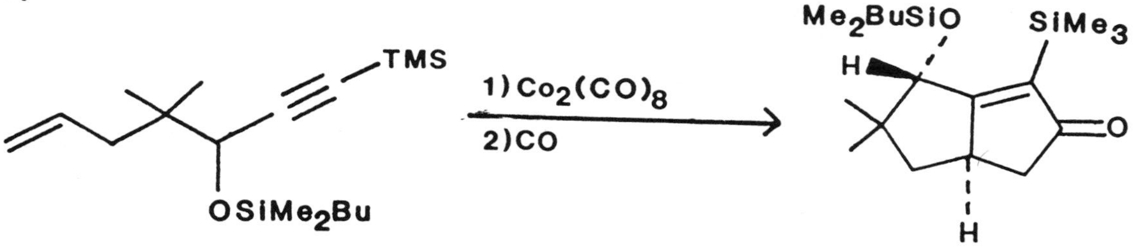

1) $Co_2(CO)_8$
2) CO

79% yield

Exon, C.; Magnus, P. **J.** **Am.** **Chem.** **Soc.** **1983**, _105_, 2477.

8. c)

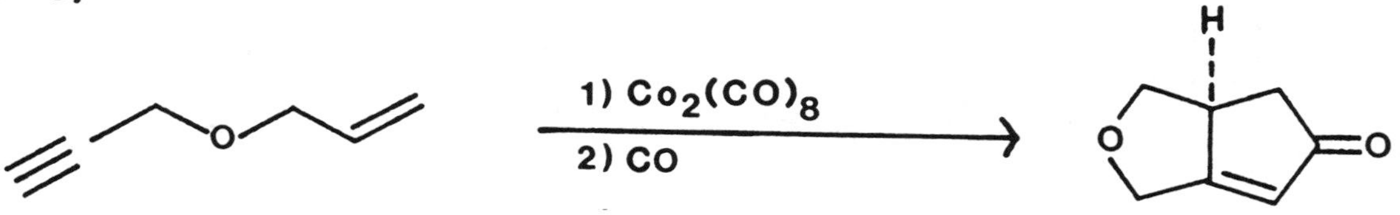

1) $Co_2(CO)_8$
2) CO

90% yield

Billington, D.C.; Willison, D. Tetrahedron Let. **1984**, _25_, 4041.

9. Complex **18** is synthesized as follows:

$$(CO)_5Cr=C=O \longrightarrow (CO)_5Cr=\underset{Me}{\overset{OLi}{<}} \xrightarrow{Me_3\overset{+}{O} B\overset{-}{F_4}} (CO)_5Cr=\underset{Me}{\overset{OMe}{<}}$$

$$\underset{Li-CH_3}{}$$

$$\underline{18}$$

Two possible mechanisms for the formation of **19** are shown in the scheme below:

9. (cont.) This reaction is significant because it readily affords β-lactams, which are structurally related to penicillin and penicillin derivatives.

McGuire, M.A.; Hegedus, L.S. _J_. _Am_. _Chem_. _Soc_. **1982**, _104_, 5538.

10. Structures for **20** and **21** are shown below:

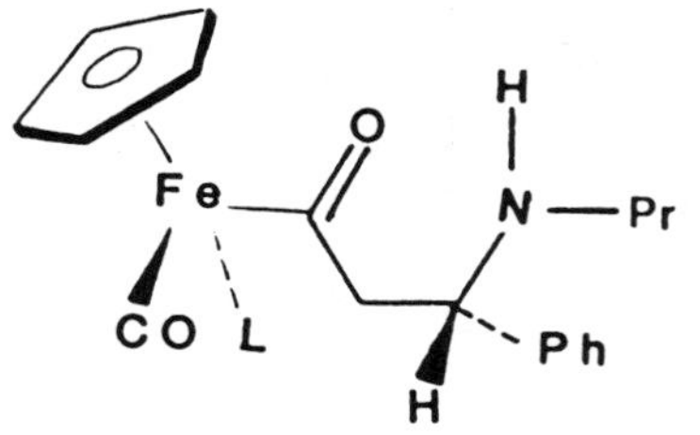

20   2 1

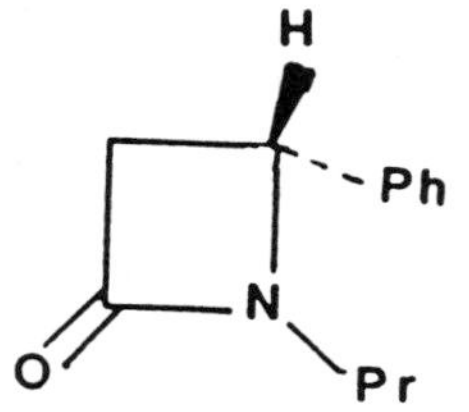

The spectral data for **20** are assigned as follows:

Spectral data for 20

IR (cm^{-1})

1918 (Fe-C≡O)

1591 (C=O)

^{1}H NMR (δ, CDCl$_3$)

0.79 (t, J=7 Hz, 3 H, $-NCH_2CH_2CH_3$)

1.36 (m, 2 H, $-NCH_2CH_2CH_3$)

2.16 (m, 2 H, $-NCH_2CH_2CH_3$)

2.84 (dd, J=18, 2 Hz, 1 H, $-COCHH'CHPh-$)

3.26 (dd, J=18, 10 Hz, 1 H, $-COCHH'CHPh-$)

3.47 (dd, J=10, 2 Hz, 1 H, $-COCHH'CHPh-$)
4.41 (d, J=1 Hz, 5 H, Cp)

7.07-7.73 (m, 15 H, PPh$_3$)

Liebeskind, L.S.; Welker, M.E.; Goedken, V. _J_. _Am_. _Chem_. _Soc_. **1984**, _106_, 441.

11. Compound **22** is shown below:

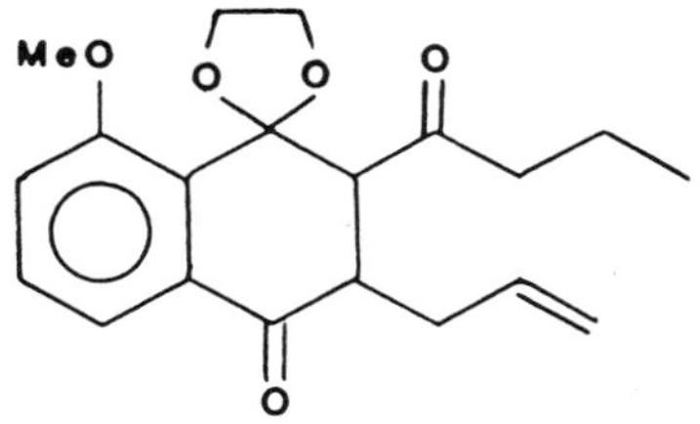

22

The proposed mechanism for the formation of **22** is shown on the following page.

11. (cont.)

$$Pr\text{-}Li \ + \ Ni(CO)_4 \longrightarrow Pr\text{-}\overset{\displaystyle O}{\overset{\|}{C}}\text{-}\bar{N}i(CO)_3$$

$$\underline{\mathbf{22}} \longleftarrow$$

Semmelhack, M.F.; Keller, L.; Sato, T.; Spiess, E. _J._ _Org._ _Chem._
1982, _47_, 4382.

13

Additional Problems

QUESTIONS

1. Propose a mechanism for the catalytic reaction shown below, which is consistent with the following observations.

$$H-\overset{\overset{\displaystyle O}{\|}}{C}-OCH_3 \longrightarrow CH_3-\overset{\overset{\displaystyle O}{\|}}{C}-OH$$

OBSERVATIONS

1) The catalyst is $[Ir(COD)Cl]_2$.

2) A carboxylic acid must be used as the solvent.

3) Methyl iodide must be added to the reaction or be generated by the reaction <u>before</u> the "isomerization" will occur.

2. A 50/50 mixture of threo- and erythro-PhCHDCHDI was allowed to react with $CpW(CO)_3^-$ to give two isomeric compounds, **3a** and **3b**.

$$\underset{\textbf{1}}{PhCHDCHDI} \quad + \quad \underset{\textbf{2}}{CpW(CO)_3^-} \quad \longrightarrow \quad \underset{\textbf{3}}{Cp(CO)_3W-CHDCHDPh}$$

$^1H\{^2H\}$ NMR (δ, $CDCl_3$) of **1**

3.08 (2 d, J=9.4, 6.0 Hz)
3.32 (2 d, J=9.4, 6.0 Hz)
plus phenyl resonanes

$^1H\{^2H\}$ NMR (δ, $CDCl_3$) of **3**

1.71 (2 d, J=13.1, 4.6 Hz)
2.77 (2 d, J=13.1, 4.6 Hz)
plus phenyl resonances

On the other hand, reaction of **2** with 100% erythro-PhCHDCHDOTs generates only one product, **3a**.

$$\underset{\textbf{1a}}{PhCHDCHDOTs} \quad + \quad \textbf{2} \quad \longrightarrow \quad \underset{\textbf{3a}}{Cp(CO)_3W-CHDCHDPh}$$

$^1H\{^2H\}$ NMR of **3a**

1.71 (d, J=4.6 Hz)
2.77 (d, J=4.6 Hz)
plus phenyl resonances

The reactions of **3a** with SO_2 and I_2 proceed as follows:

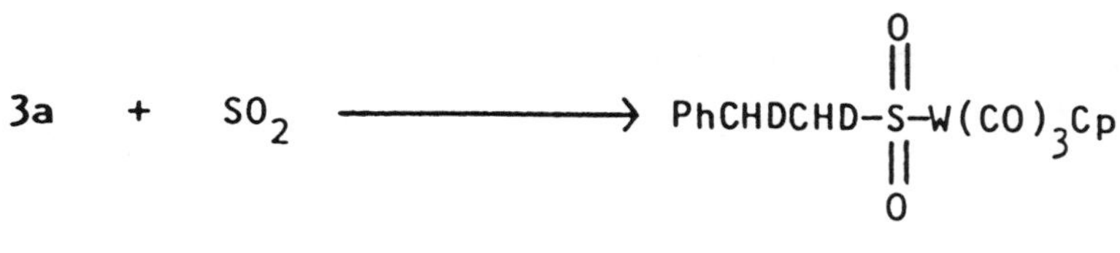

$$\textbf{3a} \quad + \quad SO_2 \quad \longrightarrow \quad PhCHDCHD-\overset{\overset{\displaystyle O}{\|}}{\underset{\underset{\displaystyle O}{\|}}{S}}-W(CO)_3Cp$$

2. (cont.)

^{1}H {^{2}H} NMR of **4**

3.10, 3.44 (AB q, J=12Hz)
plus phenyl resonances

3a + I$_2$ $\longrightarrow$ PhCHDCHDI

                              5

^{1}H {^{2}H} NMR of **5**

3.06, 3.30 (AB q, J=6 Hz)
plus phenyl resonances

a) Are compounds **3a**, **4**, and **5** the threo or the erythro isomers? Why?

b) Do the reactions to form **3a**, **4**, and **5** go with retention or inversion of configuration at the reacting carbon?

3. The following reaction produces **6** in high yield. Propose a sequence of reaction steps which shows how the product is formed.

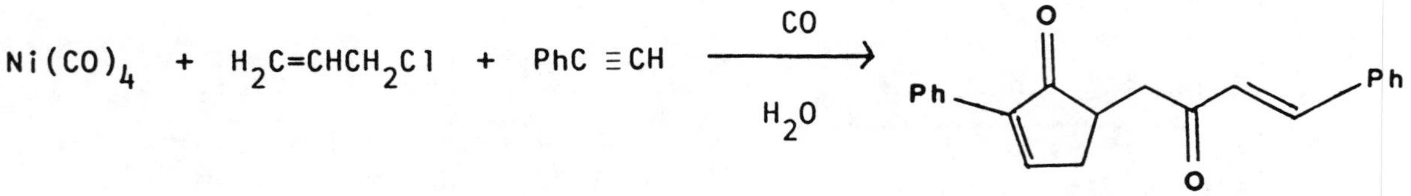

Ni(CO)$_4$ + H$_2$C=CHCH$_2$Cl + PhC≡CH $\xrightarrow[\text{H}_2\text{O}]{\text{CO}}$

4. Predict the product and propose a mechanism for the following reaction.

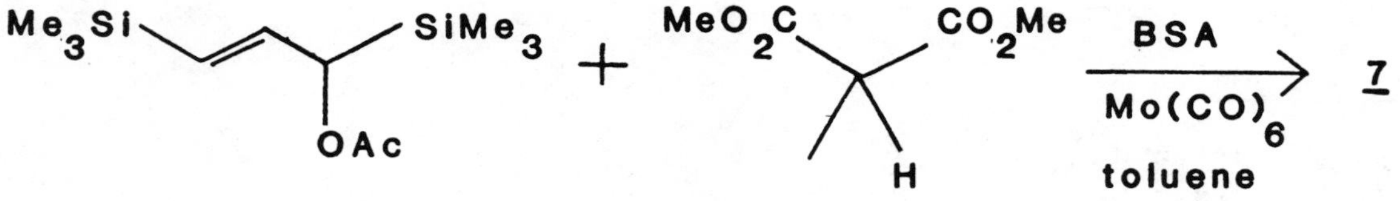

HINT
BSA [O,N-bis(trimethylsilyl)acetamide] acts as a base.

5. Give structures for compounds **10**, **11**, **12**, **13**, and **14**, and propose a mechanism for the formation of **14** from **13**.

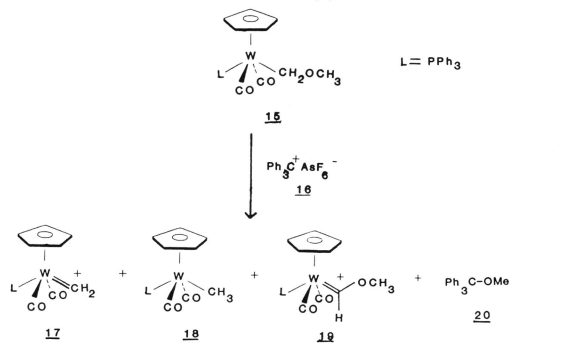

^{1}H NMR data for **14** (δ, $CDCl_3$)

1.01 (t, J=7 Hz, 3 H)
1.54 (s, 3 H)
1.89 (q, J=7 Hz, 2 H)
3.65 (s, 1 H)
3.72 (s, 1 H)
4.77 (s, 5 H)

6. When complex **15** is allowed to react with a source of triphenylmethyl cation (**16**), the following products result:

6. (cont.)
HINT
18 and **19** are formed in a 1:1 ratio.

a) Explain the formation of compounds **17**, **18**, **19**, and **20**.

b) Design an experimental procedure that will cleanly generate **17**, i.e., a procedure that will not form any **18** or **19**.

7. Propose a mechanism for the following cyclization. In addition, predict the major side product from this reaction.

Your mechanism must be consistent with the following experimental observations:

1) Radical scavengers have no effect on this reaction.

2) It is intramolecular.

3) If the aldehydic proton is replaced by deuterium, then cyclopentanone-3-d_1 is formed.

8. Nickelocene (Cp_2Ni) reacts in a Diels-Alder fashion with dimethyl acetylenedicarboxylate to give the <u>syn</u> isomer **25**. It has been found that $CpFe(CO)_2$(cyclopentadienyl), **26**, also undergoes a Diels-Alder reaction, with the major product being the <u>anti</u> isomer, **27**. Suggest a mechanism(s) which explains the difference in the product stereochemistry for these two seemingly related reactions.

8. (cont.)

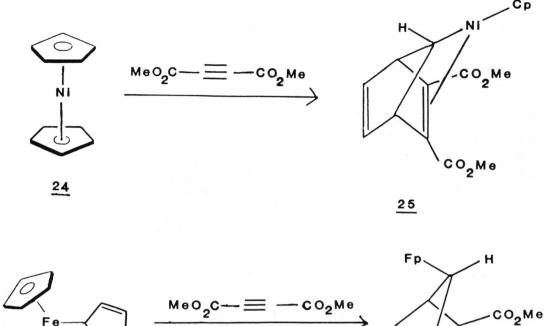

24   25

26   27

9. Propose structures for compounds 28, 29, 30, and 31.

28 + 29

30 + 31

9. (cont.)

Spectral data for **28**

IR (neat, cm^{-1})
2010

^{1}H NMR (δ, benzene-d$_6$)
2.24 (m, 2 H)
2.44 (m, 2 H)
5.02 (m, 4 H)
5.68 (m, 1 H)
5.92 (m, 1 H)

Spectral data for **29**

IR (KBr, cm^{-1})
2040, 1675, 1603

^{1}H NMR (δ, benzene-d$_6$)
1.18 (dt, J=7.5, 1.5 Hz, 3 H)
2.30 (m, 2 H)
4.84 (t, J=2.5 Hz, 2 H)
5.48 (t, J=2.3 Hz, 2 H)
5.70 (m, 1 H)

Spectral data for **30**

IR (KBr, cm^{-1})
2005, 1910

^{1}H NMR (δ, CDCl$_3$)
2.32 (d, J=7.5 Hz, 2 H)
2.74 (d, J=5.0 Hz, 2 H)
5.03 (t, J=2.2 Hz, 2 H)
5.54 (t, J=2.2 Hz, 2 H)
6.40 (m, 2 H)

Spectral data for **31**

IR (KBr, cm^{-1})
2015, 1915, 1645

^{1}H NMR (δ, CDCl$_3$)
1.68 (dt, J=8.0, 1.5 Hz, 3 H)
3.04 (m, 2 H)
5.12 (t, J=2.0 Hz, 2 H)
5.78 (m, 1 H)
5.85 (t, J=2.0 Hz, 2 H)

10. a) Propose structures for compounds **33** and **34**.

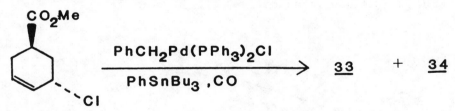

$$\text{32} \xrightarrow[\text{PhSnBu}_3, \text{CO}]{\text{PhCH}_2\text{Pd(PPh}_3)_2\text{Cl}} \quad \underline{\textbf{33}} \quad + \quad \underline{\textbf{34}}$$

32

Spectral data for **33**

IR (neat, cm^{-1}): 1740

Spectral data for **34**

IR (neat, cm^{-1}): 1740, 1680

10. (cont.)

^{1}H NMR (δ, $CDCl_3$) for **33**

1.58–1.68 (ddd, 1 H)
2.11–2.20 (ddd, 1 H)
2.20–2.40 (m, 2 H)
2.46–2.65 (dddd, 1 H)
3.40–3.60 (m, 1 H)
3.67 (s, 3 H)
5.70–6.00 (m, 2 H)
7.15–7.35 (s, 5 H)

^{1}H NMR (δ, $CDCl_3$) for **34**

1.87–1.97 (ddd, 1 H)
2.00–2.07 (ddd, 1 H)
2.25–2.45 (m, 2 H)
2.92–2.98 (dddd, 1 H)
3.70 (s, 3 H)
4.15–4.25 (m, 1 H)
5.70–5.95 (m, 2 H)
7.40–7.60 (m, 3 H)
7.90–8.00 (m, 2 H)

b) Given the fact that $PhCH_2Pd(PPh_3)_2Cl$ is present in catalytic amounts and that in solution it will generate a Pd(0) species, propose a mechanism for the formation of **33** and **34**.

11. Propose a mechanism for the following reaction.

$$(CO)_4Fe\text{-}SiMe_3^- \;+\; H_3C\text{-}\overset{\overset{\text{O}}{\|}}{C}\text{-}Br \;\xrightarrow{0\,^\circ C}\; H_2C\text{=}CHOSiMe_3 \;+\; Fe_3(CO)_{12}$$

HINT

If the reaction is carried out at $-50\,^\circ$C, an intermediate with the following spectral data is observed.

IR (cm^{-1})	^{1}H NMR (δ, CD_2Cl_2)	^{13}C NMR (δ, CD_2Cl_2)
2058 (m)	0.45 (s, 9 H)	0.2
1988 (m)	2.98 (s, 3 H)	51.0
1963 (s)		215.2
1954 (s)		341.3
1944 (sh)		

12. Propose structures for compounds **35** and **36**.

$$Cp_2TiCl_2 \;+\; (CH_3)_2C\text{=}C\overset{OLi}{\underset{H}{\diagdown}} \;\longrightarrow\; \mathbf{35}$$

$$Cp_2Zr\overset{Cl}{\underset{CH_3}{\diagup}} \;+\; (CH_3)_2C\text{=}C\overset{OLi}{\underset{H}{\diagdown}} \;\longrightarrow\; \mathbf{36}$$

ANSWERS

1. RCOOH is any solvent carboxylic acid.

1) The first step is a transesterification.

RCOOH   +   HCOOCH$_3$ $\longrightarrow$ RCOOCH$_3$ +   HCOOH

2) The second step is an oxidative-addition to a solvated monomeric iridium iodide complex derived from the iridium dimer and iodide added to or formed in the reaction solution.

$$
\text{L-Ir-I} \;+\; \text{CH}_3\text{I} \;\longrightarrow\;
\begin{array}{c}
\text{CH}_3 \\
| \\
\text{L-Ir-I} \\
| \\
\text{I}
\end{array}
$$

3) The next step is a ligand substitution reaction.

$$
\begin{array}{c}
\text{CH}_3 \\
| \\
\text{L-Ir-I} \\
| \\
\text{I}
\end{array}
\;+\; \text{HCOOH} \;\longrightarrow\;
\begin{array}{c}
\text{CH}_3 \\
| \\
\text{L-Ir-O-CHO} \\
| \\
\text{I}
\end{array}
\;+\; \text{HI}
$$

4) The fourth step is a hydrogen migration from carbon to the metal.

5) This step is followed by a methyl migration from the metal to the carbon.

1. (cont.)
6) The product-forming step is a reductive elimination.

$$
\begin{array}{c}
L \\
| \\
I-Ir-O-COCH_3 \\
| \\
H
\end{array}
\quad\longrightarrow\quad L-Ir-I \;+\; H_3CCOOH
$$

7) The final step is the regeneration of methyl iodide from the ester of the solvent and the HI generated in step 3.

$$RCOOCH_3 \;+\; HI \;\longrightarrow\; RCOOH \;+\; CH_3I$$

Pruett, R.L.; Kacmarcik, R.T. Organomet. **1982**, <u>1</u>, 1693.

2. The reaction of **1** and **2** must give a 50/50 mixture of the erythro and threo products. (In the drawing below, W represents $CpW(CO)_3$.)

erythro   threo

The two H-H coupling constants are 4.6 and 13.1 Hz. The dihedral angle betweeen the hydrogens in the erythro isomer is $180°$, while in the threo isomer it is $60°$. Thus, by the Karplus relationship, the threo isomer must have the smaller J value, and the erythro isomer, the larger value. Therefore, $J_{erythro}$ = 13.1 Hz and J_{threo} = 4.6 Hz.

From these values, it may be concluded that the reaction of **1a** and **2** gives only the threo product, i.e., complete inversion takes place in this reaction.

1a   **3a**

2. (cont.)
The reaction of **3a** with SO_2 generates **4**, which has a coupling constant of 12 Hz. This value indicates that **4** is the erythro isomer, and thus, the SO_2 must insert into the tungsten—carbon bond with inversion at the carbon.

The reaction of **3a** with I_2 generates **5**, which has a coupling constant of 6 Hz. This is the same as the smaller J value reported for the authentic threo/erythro PhCHDCHDI, and therefore, **5** must be the threo isomer. Thus, retention of configuration has occurred in the iodination reaction.

Su, S.-C.H.; Wojcicki, A. Organomet. **1983**, **2**, 1296.

3. At first you may think that this problem is very tricky; however, it consists only of reactions which have been encountered before——oxidative-addition, olefin insertion, carbonyl insertion, and reductive-elimination.

First consider the origin of each piece of the product. Carbon 1 is derived from a CO, carbons 2 and 3 from the acetylene, carbons 4, 5, and 6 from the allyl chloride, carbon 7 from a CO, and carbons 8 and 9 are derived from another acetylene.

6

The proposed mechanism is:

(mechanism continued on following page)

3. (cont.)

Heck, R.F. Accts. Chem. Res. 1969, 2, 10.

4. The product and proposed mechanism are shown below.

$$Mo(CO)_6 \xrightarrow{\Delta} Mo(CO)_5 + CO$$

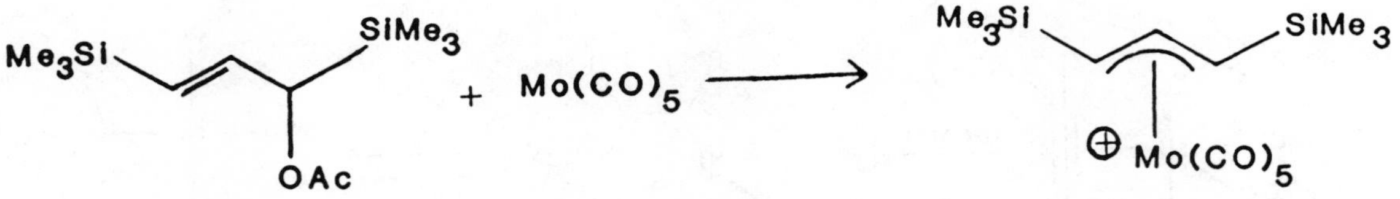

8

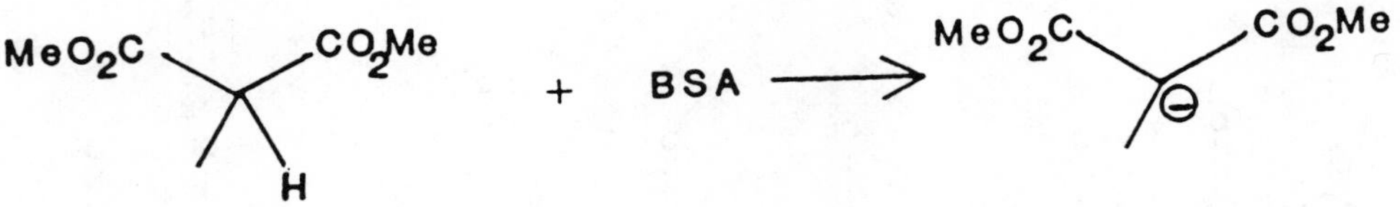

9

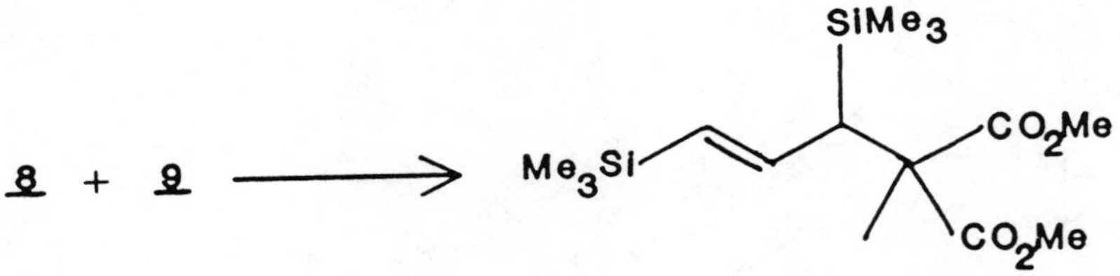

8 + 9 $\longrightarrow$

7

Trost, B.M.; Brandi, A. J. Org. Chem. 1984, 49, 4811.

5. The structures of compounds **10, 11, 12, 13,** and **14** are:

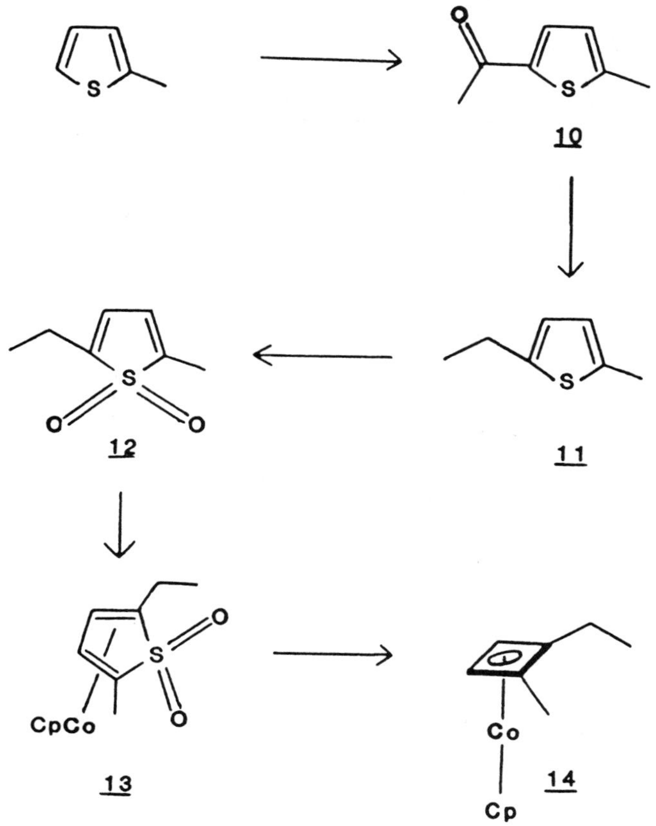

The authors propose four different mechanisms for this reaction. Based upon other work, they prefer the following mechanism:

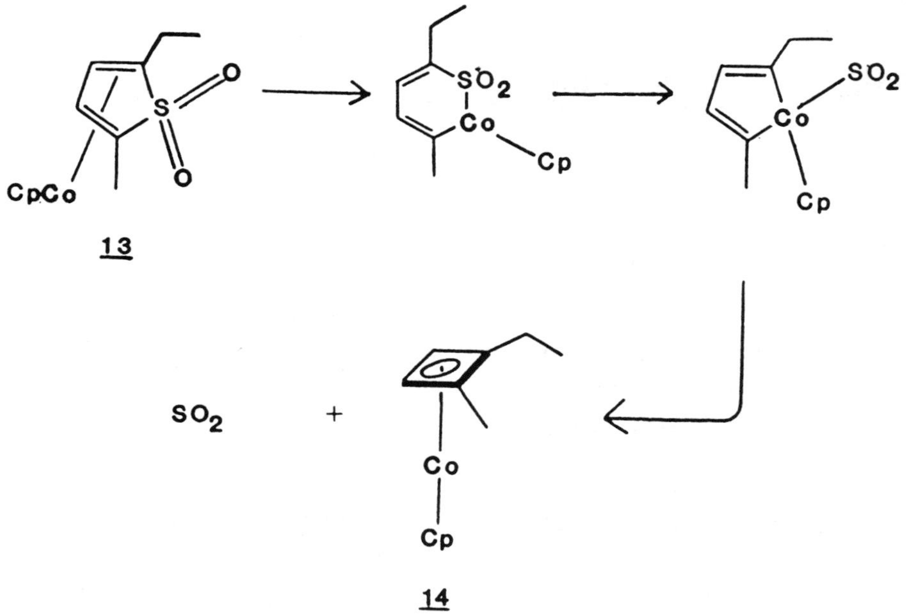

Drage, J.S.; Vollhardt, K.P.C. Organomet. 1985, 4, 389.

6. Compound **17** is formed in the following manner:

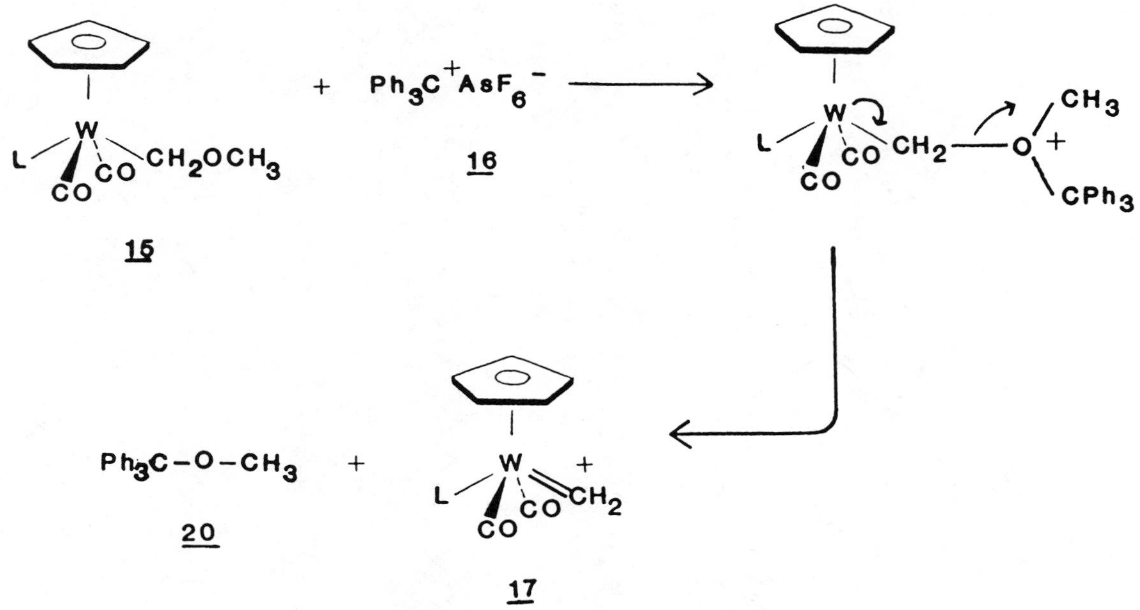

$+$ $Ph_3C^+AsF_6^-$

16

$Ph_3C-O-CH_3$ $+$

20

17

However, **17** is a highly reactive hydride abstractor and can react with **15** to give **18** and **19** in a 1:1 ratio.

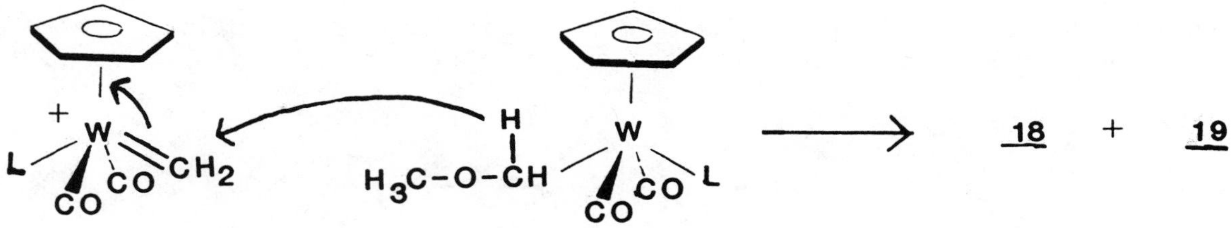

$\longrightarrow$ **18** $+$ **19**

Generation of **17** without contamination by **18** and **19** can be achieved by the slow addition of **15** to a solution of **16** or by reaction of **15** with a more reactive alkoxide abstractor such as $Me_3SiOSO_2CF_3$. The ultimate goal is to generate **17** with little **15** present to react with it.

Kegley, S.E.; Brookhart, M.; Husk, G.R. Organomet. **1982**, _1_, 760.

Davies, S.G.; Maberly, T.R. _J. Organomet. Chem._ **1985**, _296_, C37.

7. The three steps of this mechanism are an oxidative-addition, an olefin insertion, and a reductive-elimination.

The major side product arises from a decarbonylation of the organic molecule by a reverse carbonyl insertion reaction.

Campbell, R.E.; Lochow, C.F.; Vora, K.P.; Miller, R.G. J. Am. Chem. Soc. **1980,** _102_, 5824.

Milstein, D. J. Chem. Soc., Chem. Comm. **1982,** 1357.

8. In the nickel example, the dienophile coordinates with the sixteen-electron nickel complex prior to the Diels-Alder reaction. This forces the <u>syn</u> addition as shown:

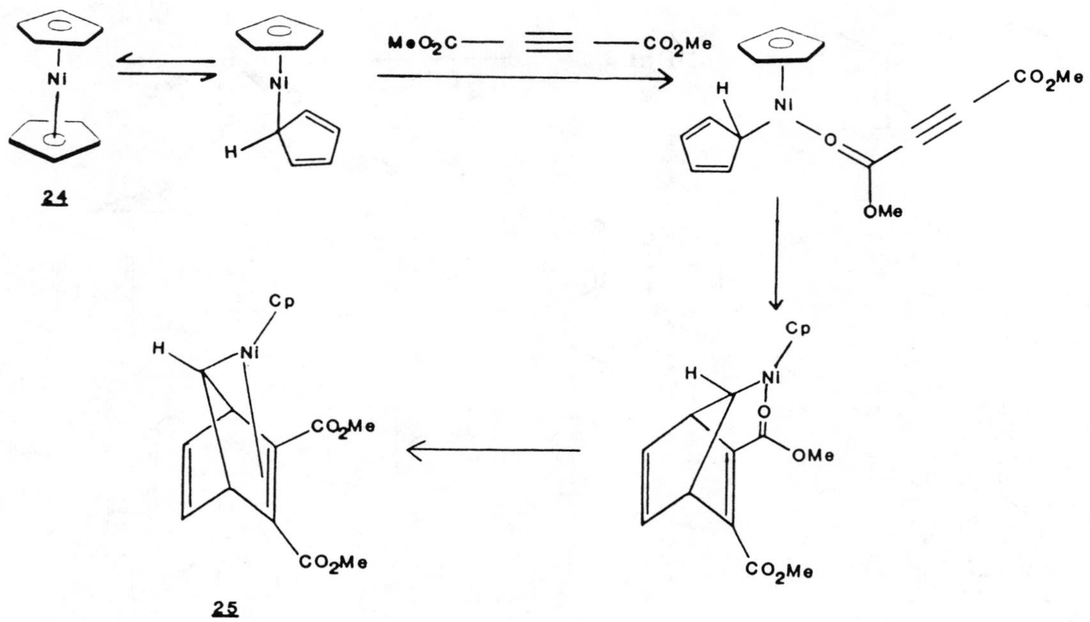

On the other hand, complex **26** has eighteen electrons around the metal, and therefore, prior coordination of the dienophile is not possible. This results in attack of the dienophile on the olefin from the less sterically crowded side of the diene, i.e., <u>anti</u> to the metal.

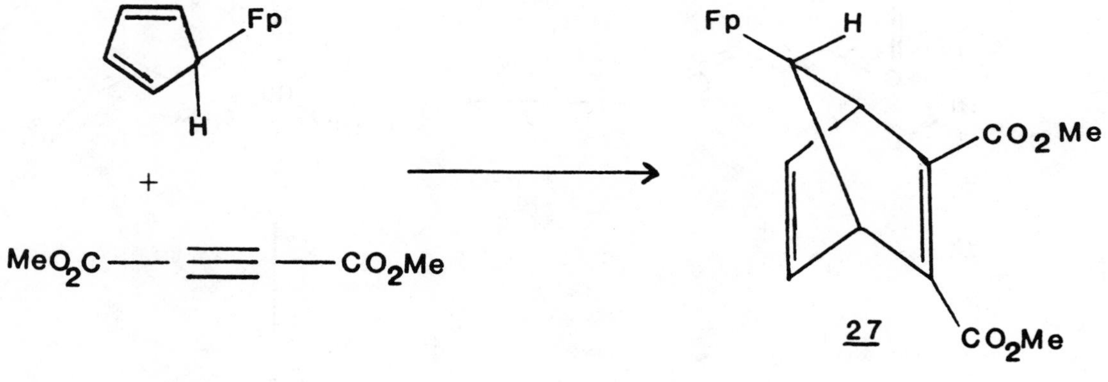

NOTE

Compound **26** also reacts with maleic anhydride, dimethyl fumarate, and chloroacrylonitrile, but not with methyl acrylate or dimethyl maleate.

8. (cont.)

In addition, the iron may be oxidatively removed from the product with Ce(IV) in methyl alcohol to generate a carboxylic derivative.

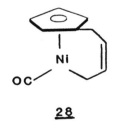

Wright, M.E. Organomet. **1983**, 2, 558.

9.

In all cases, it is the 1-3 bond of the starting material which is broken by the transition metal.

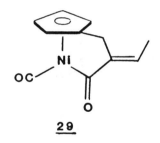

28

29

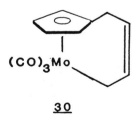

30

31

Eilbracht, P. J. Organomet. Chem. **1976**, 120, C37; Chem. Ber. **1976**, 109, 1429.

Eilbracht, P.; Mayser, U.; Tiedtke, G. Chem. Ber. **1980**, 113, 1420.

10. **a)** and **b)** The tin complex is the nucleophile and the allyl-palladium complex is the electrophile.

Sheffy, F.K.; Godschalx, J.P.; Stille, J.K. J. Am. Chem. Soc. **1984**, 106, 4833.

11. The proposed mechanism is shown on the following page. The first step is a nucleophilic attack to give an acyl silyl iron complex which rapidly rearranges to the heterocarbene complex, **37**. It is **37** which is detected at low temperature. At room temperature, this carbene complex rearranges to the olefin complex, followed by dissociation of the free olefin.

11. (cont.)

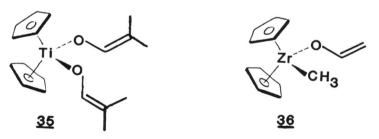

Brinkman, K.C.; Blakeney, A.J.; Krone-Schmidt, W.; Gladysz, J.A. Organomet. **1984**, <u>3</u>, 1325.

12. The enolate is a strong nucleophile and attacks the electrophilic metal complex with subsequent loss of chloride.

Curtis, M.D.; Thanedar, S.; Butler, W.M. Organomet. **1984**, <u>3</u>, 1855.

COMPOUND INDEX

A note about the compound index. The compounds are listed according to metal in a logical order, but not strictly by molecular weight. No simple organic molecules are included. The simplest mononuclear compounds (no unsaturated organic ligands of hapticity greater than two) are listed first, followed by other mononuclear complexes in order of increasing hapticity to unsaturated organic ligands. Next are polynuclear complexes, listed in order of the number of metals contained in the complex. Within a given subset (e.g., CpFe-complexes), the neutral complexes are listed first, followed by the charged species.

This Book Belongs To

· ·

The Lost Elf

Written By Victoria Cornelius

To my children, Charlie, Harry, Archie and Sophie. Buddy is lucky he found you!
And to number 5, who will be meeting Buddy for the first time this year
and his love for Christmas will begin.

Illustrations by Tig Sutton

Published by The Playwrite Group Plc, Oxted, RH8 0QA
www.playwritegroup.com
Copyright© 2016 The playwrite Group Plc
ISBN 978-1-5262-0384-7

A catalogue record for this book is available from the British Library

Each year at the beginning of December,
Father Christmas sends his elf helpers all around the world
to live with families during the build up to Christmas.

An elf's job is to report back to Father Christmas
to help him write his naughty and nice list!
But they also have lots of fun - some elves bring treats for the family,
some elves get up to mischief around the house.
Children love their elf visitors!

Buddy the elf was looking forward to returning to his family,
but when Father Christmas' sleigh dropped him off,
the house was empty, dark and cold!
The family had moved away and forgotten to tell
Father Christmas where they were going.

Facing a long, cold December alone made Buddy sad.
So after a lonely night in the dark and empty house,
he decided to pack up his belongings and try to find
a new family to spend Christmas with.

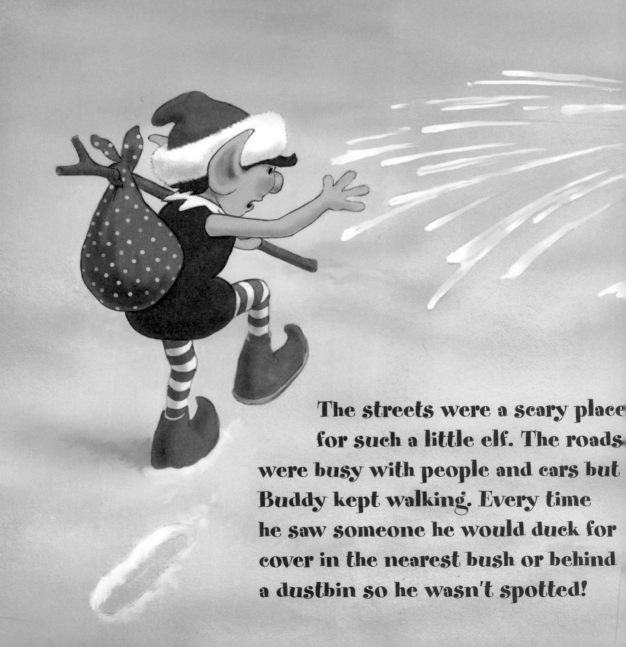

The streets were a scary place
for such a little elf. The roads
were busy with people and cars but
Buddy kept walking. Every time
he saw someone he would duck for
cover in the nearest bush or behind
a dustbin so he wasn't spotted!

He walked for what felt like miles and
miles until he finally realised
          .....he was totally lost!
As it started getting darker and colder
Buddy thought he had better find shelter
for the night.

Just as Buddy was about to give up hope a giant snowball landed on the back of his head! He turned around to see a group of Elves playing in the snow.

Overjoyed at seeing some friendly faces he hurried over and explained that he was lost and needed some help.

Luckily, one of the elves knew of a family close by
who had never had an elf visit them.
He gave a grateful Buddy the directions and waved
their friend goodbye!
With a skip in his step Buddy hurried to find the
large grey house and turning the corner he heard
the sound of children's laughter not too far away.

He walked up the garden path
to investigate and the giggles
grew louder. Using a bush to
climb, Buddy reached the
window and peered through.
Inside were four children sitting
in front of a beautiful Christmas
tree, laughing as the eldest
boy read a bedtime story.

The family looked perfect to Buddy, so happy and festive.
As the story finished their Mummy and Daddy took the children up to bed.
Buddy was looking forward to hopefully becoming part of their family.

When all the children and grown ups were in bed and asleep, he used some Christmas magic to drop down the chimney, into the living room.

In the morning, the 4 children woke to a very curious sight -
a little elf asleep underneath their Christmas tree!!
A note beside him read:

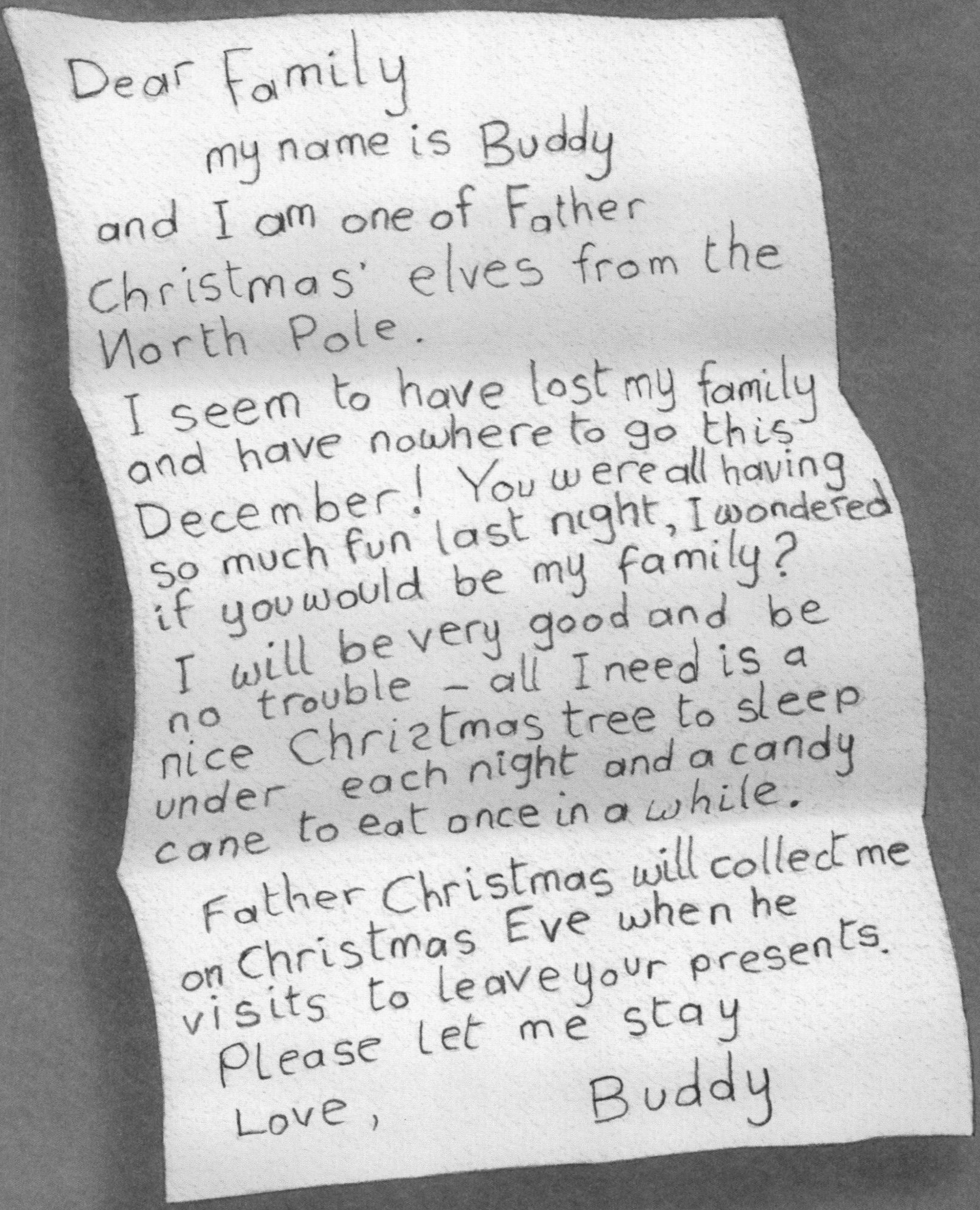

Dear Family
         my name is Buddy
and I am one of Father
Christmas' elves from the
North Pole.
   I seem to have lost my family
and have nowhere to go this
December! You were all having
so much fun last night, I wondered
if you would be my family?
   I will be very good and be
no trouble - all I need is a
nice Christmas tree to sleep
under each night and a candy
cane to eat once in a while.
   Father Christmas will collect me
on Christmas Eve when he
visits to leave your presents.
   Please let me stay
   Love,   Buddy

The children were so excited at the thought of having Buddy as their elf!

Their Mummy and Daddy thought it was a lovely idea and said of course he could stay.

Buddy brought even more magic into their house
throughout December and on Christmas Eve,
everyone was sad knowing that he would be leaving them.
But Buddy knew this family would be there waiting for him
next Christmas and he was already counting down the days!

Merry Christmas.
See you next year!

The End